OTHER TITLES OF INTEREST FROM ST. LUCIE PRESS

Trace Substances in Environmental Health (Vols. 23–25)

Contamination of Groundwaters

Lead in Soil: Recommended Guidelines

Metal Compounds in Environment and Life: Interrelation Between Chemistry and Biology

Emergency Response and Hazardous Chemical Management: Principles and Practices

Geochemistry and Health

Naturally Occurring Radioactive Material: Principles and Practices

Health Effects from Petroleum Hydrocarbon Exposure in Confined Spaces: Principles and Case Studies

Chemistry and Health Effects of Hazardous Chemicals

Healthcare Safety Management

For more information about these titles call, fax or write:

St. Lucie Press
2000 Corporate Blvd., N.W.
Boca Raton, FL 33431-9868

TEL (561) 994-0555 • (800) 272-7737
FAX (800) 374-3401
E-MAIL information@slpress.com
WEB SITE http://www.slpress.com

S^t_L

The Handbook of
Trace
Elements

The Handbook of
Trace
Elements

István Pais
J. Benton Jones, Jr.

S^t_L

St. Lucie Press
Boca Raton, Florida

Printed and bound in the U.S.A. Printed on acid-free paper.
10 9 8 7 6 5 4 3 2 1

ISBN 1-884015-34-4

Phone: (561) 994-0555
E-mail: information@slpress.com
Web site: http://www.slpress.com

S_{L}^{t}

Published by
St. Lucie Press
2000 Corporate Blvd., N.W.
Boca Raton, FL 33431-9868

TABLE OF CONTENTS

PREFACE

This book summarizes, by element, 41 elements that are found in the environment (earth's crust, soil, water, plants, animals, and man) in relatively low (<0.1%) concentrations. These elements are generally referred to as trace elements or, for the seven trace elements essential to plants, micronutrients. In addition, there is a chapter on the rare earth elements.

These elements are of increasing interest to ecologists and plant, animal, and human physiologists because their presence in the environment can have profound effects on crop production as well as the well-being of animals and man. Recent discoveries suggest that a number of the trace elements can have beneficial effects on the health of animals and man at concentrations of 1 mg/kg or less, whereas higher concentrations are frequently detrimental.

Some of the trace elements are found concentrated in both naturally occurring and man-made substances. Most industrial waste products, such as sewage sludge and animal wastes, contain sizeable quantities of the trace elements, making their disposal a significant ecological problem. Some manufacturing processes result in the concentration of various trace elements in both finished and waste products. Also, there are naturally occurring processes that concentrate certain trace elements in various sectors of the environment.

Lack of or poor growth of plants as well as animal and human health problems are being uncovered and found to stem from the lack of a normal distribution of some trace elements within the environment. Such conditions require the supplementation of soils and diets to overcome the occurrence of deficiencies. Deficiencies can also occur due to certain environmental conditions, such as very acid or alkaline soils, which will influence the availability of an essential trace element to plants. Toxicities also can occur if availability is markedly increased with changing soil or dietary conditions. The presence as well as lack of certain trace elements in drinking and irrigation water and the trace element content of ingested dust can have significant effects on the health of animals and man.

Of the 105 known elements, 29 are believed to be essential for the growth and development of living organisms. Of these 29 elements, 18 are classed as trace elements, with 10 known as transition metals. Due to their biological activity, about one-half of the essential trace elements function as metallo-enzymes; this includes the elements cobalt, copper, iron, manganese, molybdenum, nickel, selenium, and zinc. Other trace elements—whose exact function of essentiality is not fully known—are metals and non-metals: arsenic, boron, cadmium, chromium, fluorine, lead, lithium, silicon, tin, and vanadium.

A number of the trace elements have been clearly identified as being essential for the normal growth, development, and well-being of plants, animals, and man. Seven trace elements, referred to as micronutrients, have been identified as essential for plants: boron, chlorine, copper, iron, manganese, molybdenum, and zinc. There is increasing evidence which suggests that nickel should be added to the list of essential micronutrients for plants. For the well-being of animals and man, 12 trace elements have been identified as being essential: arsenic, chlorine (also considered a macroelement due to its relatively high requirement for sufficiency), chromium, copper, fluorine, iodine, iron, manganese, molybdenum, nickel, selenium, and vanadium.

Other trace elements have been found to promote various physiological processes that occur in plants, animals, and man, suggesting that an additional category should be established which would identify these elements are being beneficial. No such category has been accepted, although many physiologists are well aware of the promotive influence that a number of these trace elements have.

We have drawn on a wide range of sources in both the current and past published literature in order to assemble the data given in this book for each trace element. In assembling these data, inconsistencies were frequently encountered in terms of the concentration found in various substances; therefore, we selected data that were consistent if found in two or more sources. In those cases where only one source was available, the data from that source were used.

Our major objective is to assemble as much fundamental and factual information as possible on the trace elements, in order to make this book an easy-to-use reference that provides data which currently exist for 41 trace elements (plus the rare earths).

István Pais
J. Benton Jones, Jr.

ABOUT
THE AUTHORS

István Pais is Professor Emeritus at the University of Horticulture and Food Industry, Budapest, Hungary. He retired from the university in 1994 after completing 30 years of service plus 20 years as Assistant, First Assistant, and Associate Professor of Inorganic and Analytical Chemistry at the Pázmány (Eötvös from 1950) University, Budapest.

Dr. Pais received his Master of Science degree in chemistry from the Pázmány University in 1945, a diploma as secondary school teacher in chemistry and physics in 1946, and a Pd.D. degree in 1947 from the same university. In 1976, he received the degree "Doctor Sciences" from the Hungarian Academy of Sciences.

The author of over 200 scientific articles and 10 book chapters, Dr. Pais has also written 12 Hungarian textbooks and was the editor of 9 Proceedings of the International Trace Element Symposium. He was a member of the Editorial Board of *Communications in Soil Science and Plant Analysis* for 20 years and since 1983 has been on the Editorial Board of the *Journal of Plant Nutrition*. Since 1947, Dr. Pais has been a member of the Hungarian Chemical Society and the Hungarian Society for Dissemination of Sciences, in which he has served as the vice-chairman of the Section of Chemistry since 1961.

Dr. Pais organized the Trace Element Committee of the Hungarian Academy of Sciences in 1981 and was re-elected as President of the committee to serve until 1999. He has also worked for eight years as a member of the Chemistry Teaching Committee of IUPAC and the Federation of European Chemical Societies.

Dr. Pais has traveled extensively as either a participant or invited lecturer at conferences held in Poland, Czechoslovakia, Germany, the Soviet Union, the United Kingdom, Belgium, the Netherlands, Sweden, Finland, France, Spain, Switzerland, Austria, Italy, the United States, Japan, China, India, and the Philippines.

The recipient of over ten awards in his native country, Dr. Pais was decorated with the highest Hungarian award, the Széchenyi Prize for Science and Education. He is listed in several Hungarian and international *Who's Who,* as well as a number of other similar biographical listings.

J. Benton Jones, Jr. is vice president of Micro-Macro International, an analytical laboratory specializing in the assay of soil, plant tissue, water, food, animal feed, and fertilizer. He is also president of his own consulting firm, Benton Laboratories; vice president of a video production company engaged in producing educational videos; and president of a new company, Hydro-Systems, Inc., which manufactures hydroponic growing systems.

Dr. Jones is Professor Emeritus at the University of Georgia. He retired from the university in 1989 after having completed 21 years of service plus 10 years as Professor of Agronomy at the Ohio Agricultural Research and Development Center, Wooster.

He received his B.S. degree from the University of Illinois in 1952 in agricultural science and a M.S. degree in 1956 and a Ph.D. degree in 1959 in agronomy from the Pennsylvania State University.

Dr. Jones is the author of over 200 scientific articles and 15 book chapters, and he has written four books. He was editor of two international journals, *Communications in Soil Science and Plant Analysis* for 24 years and the *Journal of Plant Nutrition* for 19 years. Dr. Jones is secretary-treasurer of the Soil and Plant Analysis Council, a scientific society which was founded in 1969, and has been active in the Hydroponic Society of America from its inception, serving on its board of directors for five years.

He has traveled extensively with consultancies in the Soviet Union, China, Taiwan, South Korea, Saudi Arabia, Egypt, Costa Rica, Cape Verde, India, Hungary, Kuwait, and Indonesia.

Dr. Jones has received many awards and recognition for his service to the science of soil testing and plant analysis. He is a certified soil and plant scientist under the ARPACS program of the American Society of Agronomy. He is a Fellow of the American Association for the Advancement of Science, Fellow of the American Society of Agronomy, and Fellow of the Soil Science Society of America. An award in his honor, The J. Benton Jones, Jr. Award, established in 1989 by the Soil and Plant Analysis Council, has been given to three international soil scientists, one in each of the years 1991, 1993, and 1995. Dr. Jones received an Honorary Doctor's Degree from the University of Horticulture, Budapest, Hungary, and is a member of three honorary societies, Sigma Xi, Gamma Sigma Delta, and Phi Kappa Phi. He is listed in *Who's Who in America* as well as a number of other similar biographical listings.

1
INTRODUCTION

TERMINOLOGY

Over the past century, those elements required only in very small quantities by living organisms have been variously identified in the literature as either microelements, trace elements, or micronutrients; for the seven essential trace elements required by plants—boron, chlorine, copper, iron, manganese, molybdenum, and zinc—the proper term is *micronutrient.* Therefore, the term *trace element* would then refer to those elements found in plants at low concentrations but not yet identified as essential. In the animal and human nutrition literature, reference to those elements found in low concentrations in body tissues and fluids is normally termed trace elements, whether the element is essential or not.

The word *micronutrient* can have a meaning other than that associated with the mineral elements; substances required by plants, animals, and man, such as amino acids, unsaturated fatty acids, etc.; and organic molecules required in relatively low concentrations. However, in this text, *micronutrient* will be used exclusively to classify 7 (boron, chlorine, copper, iron, manganese, molybdenum, and zinc) of the 16 essential elements required by plants, and *trace element* will be used for the non-essential elements found in low concentrations in plants. For both animals and humans, the term *trace element* will be used to designate those elements found in low concentrations.

Webster's New World Dictionary defines trace element as "a chemical element, as iron, copper, zinc, etc., essential in plant and animal nutrition,

1

but only in minute quantities." In his *Encyclopedia of Environmental Studies,* Ashworth (1991) defines trace element or micronutrient as:

> an element required in very small amounts in the diet of a living organism. The typical dietary requirement for a trace element is less than 50 parts per million, with organisms usually showing a very limited range of tolerance to deviations from the required amount. Too much is often toxic, while too little can cause severe nutritional disorders. The metabolic use of these elements is varied, but most uses are related to the enzyme system either as a structural part of various enzymes or as catalysts for their function in the body: a few form minor but essential structural components of specialized compounds such as hemoglobin (animals) and chlorophyll (plants). Examples of trace elements include copper, selenium, manganese, cobalt, molybdenum, iodine, and iron. Examples of disorders stemming from their lack in the diet include goiter (iodine deficiency), anemia (iron, copper, or cobalt deficiency), and white-muscle disease (selenium deficiency).

Another term commonly used to refer to the essential elements is *nutrient,* which has been defined in the biological sciences as "an element required by an organism to build living tissue and maintain the chemical reactions of the life process" (Ashworth, 1991). The word nutrient may also refer to a compound, such as a protein or carbohydrate, and therefore is not useful in this text for identifying a trace element or micronutrient.

When referring to the metal trace elements, the term *heavy metal* has been frequently used to identify those elements that have a relatively high atomic weight. One definition of a heavy metal is "that metal which has a density greater than 5.0 mg/m^3, with elements such as cadmium, cobalt, copper, iron, lead, molybdenum, nickel, and zinc" considered heavy metals. Ashworth (1991) defines heavy metal as "any of a large group of metallic elements with relatively high atomic weights and similar health effects." In the definition, Ashworth (1991) also states that "the group is somewhat poorly defined" and that "the group ranges down to atomic weight 24 (chromium) and includes at least two non-metals, arsenic and selenium."

Those elements identified as trace elements or micronutrients in this text are included as such in Figure 1.1 in a Periodic Table configuration.

The chemical properties of the 41 trace elements specifically discussed in this text are given in Table 1.1.

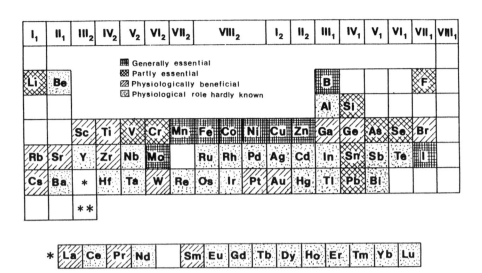

FIGURE 1.1 Periodic Table identification of the trace elements and micronutrients discussed.

TABLE 1.1 Chemical properties of the trace elements

Element	Group in Periodic Table	Atomic Number	Atomic Weight	Ions	Ionic Radius[a]	Electro-negativity[b]	Ion Potential (charge/radius)
Aluminum (Al)	IIIA	13	26.98	Al^{3+}	0.57	1.6	—
Antimony (Sb)	VA	51	121.75	Sb^{3+}	0.76	—	—
				Sb^{5+}	0.62	1.9	—
Arsenic (As)	VA	33	74.92	As^{3+}	0.58	—	—
				As^{5+}	0.46	—	—
Barium (Ba)	IIA	56	137.32	Ba^{2+}	1.43	0.9	—
Beryllium (Be)	IIA	4	9.01	Be^{2+}	0.34	1.6	—
Bismuth (Bi)	VA	83	208.98	Bi^{5+}	0.74	2.0	—
				Bi^{3+}	0.96		
Boron (B)	IIIA	5	10.81	B^{3+}	0.23	2.0	—
Cadmium (Cd)	IIB	48	122.40	Cd^{2+}	0.97	1.7	—

TABLE 1.1   Chemical properties of the trace elements (continued)

Element	Group in Periodic Table	Atomic Number	Atomic Weight	Ions	Ionic Radius[a]	Electro-negativity[b]	Ion Potential (charge/radius)
Cesium (Cs)	IA	55	132.91	Cs^+	1.65	0.8	—
Chromium (Cr)	VIB	24	52.00	Cr^{3+}	0.63	1.6	4.3
				Cr^{6+}	0.52	—	16.0
Cobalt (Co)	VIII	27	58.93	Co^{2+}	0.72	1.8	2.6
Copper (Cu)	IB	29	63.54	Cu^+	0.96	1.9	—
				Cu^{2+}	0.72	2.0	2.5
Fluorine (F)	VIIA	9	18.99	F^-	1.33	4.1	—
Gallium (Ga)	IIIA	31	69.72	Ga^{3+}	0.62	1.8	—
				Ga^+	1.13	—	—
Germanium (Ge)	IVA	32	72.61	Ge^{2+}	0.90	2.0	—
Gold (Au)	IB	79	196.97	Au^+	1.37	2.4	—
Hafnium (Hf)	IVB	72	178.49	Hf^{4+}	0.84	1.3	—
Indium (In)	IIIA	49	114.82	In^{3+}	0.92	1.8	—
				In^+	1.32	—	—
Iodine (I)	VIIA	53	126.90	I^-	2.20	2.7	—
Iron (Fe)	VIII	26	55.85	Fe^{2+}	0.82	1.8	—
				Fe^{3+}	0.67	—	—
Lead (Pb)	IVA	82	207.19	Pb^{2+}	1.20	1.8	1.9
Lithium (Li)	IA	3	6.94	Li^+	0.78	1.0	—
Manganese (Mn)	VIIB	25	54.94	Mn^{2+}	0.80	1.5	—
				Mn^{3+}	0.66	—	—
				Mn^{4+}	0.60	—	6.5
Mercury (Hg)	IIB	80	200.59	Hg^{2+}	1.10	1.9	—
Molybdenum (Mo)	VIB	42	95.94	Mo^{4+}	0.70	—	—
				Mo^{6+}	0.62	1.8	12.0
Nickel (Ni)	VIII	28	59.71	Ni^{2+}	0.69	1.8	2.6
Niobium (Nb)	VB	41	92.91	Nb^{4+}	0.74	1.6	—
Platinum (Pt)	VIII	78	195.08	Pt^{2+}	0.85	2.2	—
Rubidium (Rb)	IA	37	85.47	Rb^+	1.49	0.8	—
Selenium (Se)	VIA	34	78.96	Se^{2-}	12.001	2.4	3.7
				Se^{4+}	0.42	—	—
Silicon (Si)	IVA	14	28.09	Si^{4+}	0.26	1.9	—
				Si^{4-}	2.71	—	—
Silver (Ag)	IB	47	107.87	Ag^+	1.26	1.9	—
Strontium (Sr)	IIA	38	87.62	Sr^{2+}	1.27	0.9	—
Thallium (Tl)	IIIA	81	204.37	Tl^+	1.47	—	—
				Tl^{3+}	0.95	1.8	—

TABLE 1.1   Chemical properties of the trace elements (continued)

Element	Group in Periodic Table	Atomic Number	Atomic Weight	Ions	Ionic Radius[a]	Electro-negativity[b]	Ion Potential (charge/ radius)
Tin (Sn)	IVA	50	118.69	Sn^{2+}	0.93	1.8	1.5
				Sn^{4+}	0.71	1.9	—
Titanium (Ti)	IVB	22	47.88	Ti^{4+}	0.80	1.5	—
				Ti^{3+}	0.69	—	—
Tungsten (W)	VIB	74	183.85	W^{6+}	0.62	1.7	—
Uranium (U)	Actinide series	92	238.04	U^{4+}	0.97	—	—
				U^{6+}	0.80	1.7	—
Vanadium (V)	VB	23	50.94	V^{3+}	10.651	—	—
				V^{4+}	(0.65)	—	—
				V^{5+}	0.59	—	11.0
Zinc (Zn)	IIB	30	65.37	Zn^{2+}	0.74	1.7	2.6
Zirconium (Zr)	IVB	40	91.22	Zr^{2+}	1.09	1.3	—
				Zr^{4+}	0.87	—	—

[a] Ionic radius is for 6-coordination.
[b] Electronegativity values for elements: S = 2.5, O = 3.5, I = 2.5, Cl = 3.0, F = 4.0. From these numbers, it can be generalized that a bond between any two atoms will be largely covalent if the electronegativities are similar and mainly ionic if they are very different.

GEOGRAPHICAL CONTENT AND DISTRIBUTION

Lithosphere and Soil

The trace element content of the lithosphere and soil varies considerably from percent (silicon 22.7%) to the low parts per million range (platinum ~0.001 mg/kg). The abundance of the trace elements listed alphabetically is given in Table 1.2 and is given by descending concentration in Table 1.3.

The trace element content of soils has been reported by Temmerman et al. (1984), and Adriano (1986) has summarized the trace element content of soils and surface materials from several authors; the former is given in Table 1.4 and the summary in Table 1.5.

Additional soil trace element content data corresponding to plant trace element levels can be found in Table 1.35.

TABLE 1.2   Abundance of the trace elements in the lithosphere and soils

	Abundance (mg/kg)	
Trace Element	*Lithosphere*	*Soil*
Aluminum (Al)	82,000.0	10,000.0
Antimony (Sb)	0.2	0.9
Arsenic (As)	1.5	5.0
Barium (Ba)	500.0	100.0
Beryllium (Be)	2.6	<10.0
Bismuth (Bi)	0.048	<1.0
Boron (B)	10.0	8.0
Cadmium (Cd)	0.11	<1.0
Cesium (Cs)	3.0	—
Chromium (Cr)	100.0	<100.0
Cobalt (Co)	20.0	10.0
Copper (Cu)	50	5–20
Gallium (Ga)	18.0	1–40
Germanium (Ge)	1.8	1–10
Gold (Au)	0.0011	~0.1
Indium (In)	0.049	0.2
Iron (Fe)	45,000	50,000
Lead (Pb)	14.0	2–200
Lithium (Li)	20.0	10.0
Manganese (Mn)	950.0	200–3,000
Mercury (Hg)	0.05	0.03
Molybdenum (Mo)	1.5	0.2–5.0
Nickel (Ni)	~80.0	30–40
Platinum (Pt)	~0.001	—
Rubidium (Rb)	90.0	20–500
Selenium (Se)	0.05	0.1–2
Silicon (Si)	277,000.0	—
Silver (Ag)	0.07	<0.1
Strontium (Sr)	370.0	200
Tellurium (Te)	~0.005	—
Thallium (Tl)	0.6	—
Tin (Sn)	2.2	<1.0
Titanium (Ti)	5,600.0	1,800–3,600
Tungsten (W)	1.0	—
Vanadium (V)	160.0	100–1,000
Zinc (Zn)	75.0	10–300
Zirconium (Zr)	190.0	60–2,000

TABLE 1.3 Abundance of the trace elements by decreasing concentration in the lithosphere and soils

Trace Element	Abundance (mg/kg)	
	Lithosphere	Soil
Silicon (Si)	277,000	—
Aluminum (Al)	82,000	10,000
Iron (Fe)	45,000	38,000
Titanium (Ti)	5,600	1,800–3,600
Manganese (Mn)	950	200–3,000
Barium (Ba)	500	100.0
Strontium (Sr)	370	200
Zirconium (Zr)	190	60–2,000
Vanadium (V)	160	100–1,000
Chromium (Cr)	100	<100.0
Rubidium (Rb)	90	20–500
Nickel (Ni)	~80	30–40
Zinc (Zn)	75	10–300
Copper (Cu)	50	5–20
Cobalt (Co)	20	10.0
Lithium (Li)	20	10.0
Gallium (Ga)	18	1–40
Lead (Pb)	14	2–200
Boron (B)	10	8.0
Cesium (Cs)	3.0	—
Beryllium (Be)	2.6	<10.0
Tin (Sn)	2.2	<1.0
Germanium (Ge)	1.8	1–10
Arsenic (As)	1.5	5.0
Molybdenum (Mo)	1.5	0.2–5.0
Tungsten (W)	1.0	—
Thallium (Tl)	0.6	—
Antimony (Sb)	0.2	—
Cadmium (Cd)	0.11	<1.0
Silver (Ag)	0.07	<0.1
Selenium (Se)	0.05	0.1–2
Mercury (Hg)	0.05	0.03
Indium (In)	0.049	—
Bismuth (Bi)	0.048	<1.0
Tellurium (Te)	~0.005	—
Gold (Au)	0.0011	~0.1
Platinum (Pt)	~0.001	—

TABLE 1.4   Upper limits of "normal" concentration in non-polluted soils (in µg/g)

| Element | Quaternarian | | | Paleozoic Mesozoic |
	Sand	Sandy Loam	Loam	Loam and Clay
Antimony (Sb)	1	1	1	1
Arsenic (As)	10	15	20	30
Beryllium (Be)	0.5	1	2	3
Boron (B)	20 (20)	40 (40)	50 (60)	60 (100)
Cadmium (Cd)	1	1	1	1
Chromium (Cr)	15 (80)	30 (150)	30 (200)	80 (300)
Cobalt (Co)	5 (5)	10 (20)	15 (20)	20 (40)
Copper (Cu)	15	25	25	30
Gallium (Ga)	(10)	(20)	(20)	(35)
Lead (Pb)	50	50	50	50
Manganese (Mn)	500	800	800	2000
Mercury (Hg)	0.15	0.15	0.15	0.20
Molybdenum (Mo)	(5)	(5)	(5)	(5)
Nickel (Ni)	10 (10)	20 (20)	30 (40)	40 (80)
Selenium (Se)	1	1	1	—
Silver (Ag)	(0.5)	(0.5)	(0.5)	(0.5)
Strontium (Sr)	(100)	(200)	(200)	(300)
Thallium (Tl)	0.3	0.5	0.5	0.5
Tin (Sn)	(5)	(7)	(7)	(8)
Vanadium (V)	50 (100)	100 (150)	100 (200)	200 (300)
Zinc (Zn)	100	100	150	200

( ) = total concentration, including the mineral fraction.

Source: Temmerman et al., 1984.

The geographic distribution of the trace elements, whether naturally occurring or as a result of man's activity, poses problems of deficiency (e.g., lack of naturally occurring molybdenum in Australian soils for normal legume growth and selenium deficiency affecting human health in Finland) as well as toxicity (e.g., cadmium and zinc in sewage sludge and lead from vehicular traffic) (Welch et al., 1991).

There is also a fairly strong relationship between the elemental content in the earth's crust and that found in topsoils, as illustrated in Figure 1.2 (Kabata-Pendias and Pendias, 1984). The varying range in topsoil content versus that in the lithosphere illustrates the influence of dilution and con-

TABLE 1.5 Median or average contents reported for elements in soils and other surficial materials

	mg/kg				
Trace Element	Bowen Median	Shacklette and Boerngen Average	Vinogradov Average	Rose et al. Average	Median
Antimony (Sb)	1	0.66	—	2	—
Arsenic (As)	6	7.2	5	—	7.5
Barium (Ba)	500	580	—	—	300
Beryllium (Be)	0.3	0.92	—	6	0.5–4
Bismuth (Bi)	0.2	—	—	—	—
Boron (B)	20	33	10	—	29
Cadmium (Cd)	0.35	—	—	—	—
Cesium (Cs)	4	—	—	—	—
Chromium (Cr)	70	54	200	—	6.3
Cobalt (Co)	8	9.1	8	—	10
Copper (Cu)	30	25	20	—	15
Fluorine (F)	200	430	200	—	300
Lead (Pb)	35	19	—	—	17
Manganese (Mn)	1000	550	850	—	320
Mercury (Hg)	0.06	0.09	—	—	0.056
Molybdenum (Mo)	1.2	0.97	2	2.5	—
Nickel (Ni)	50	19	40	—	17
Selenium (Se)	0.4	0.39	0.001	—	0.3
Silver (Ag)	0.05	—	—	—	—
Strontium (Sr)	250	240	300	—	67
Thallium (Tl)	0.2	—	—	—	—
Tin (Sn)	4	1.3	—	10⁻	—
Titanium (Ti)	5000	2900	4600	—	—
Tungsten (W)	1.5	—	—	—	—
Vanadium (V)	90	80	100	—	57
Zinc (Zn)	90	60	50	—	36

Source: Adriano, 1986.

centration as a result of natural weathering. A good summary of data on the trace element content of soils published prior to 1969 is given by Swaine (1969) and is shown in Figure 1.3 from Kabata-Pendias and Pendias (1984).

The total trace element content of a soil may not relate to the effect that element would have on plant growth and development since only a portion

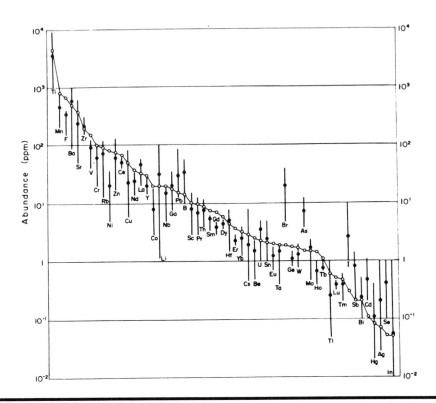

FIGURE 1.2   Trace elements in soils compared to their abundance in the lithosphere. Open circles = mean content in the lithosphere; black circles = mean content in topsoils; vertical lines = values commonly found in topsoils. (Source: Kabata-Pendias and Pendias, 1984.)

of the total is "available" to the plant which absorbs elements only from the soil solution. Therefore, knowing the composition of the soil solution would be of considerable value. The ranges in concentration in the soil solution of natural soil for nine trace elements are given in Table 1.6.

Of growing concern worldwide is the trace element content of soils as they naturally exist, as well as that from the long-term effects that fertilization and cropping have on the trace element content of cropland soils. Much of this concern has come about from the increasing use of various waste products, such as sewage sludge, applied on arable land, particularly lands being used for food crop production (Chaney, 1980; Chaney et al., 1987). Typical trace element content for sewage sludges is given in Table 1.7.

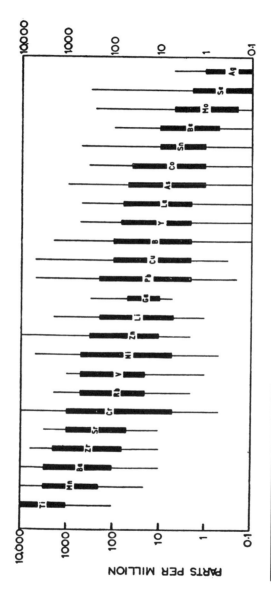

FIGURE 1.3 Total contents of trace elements in soils. (Source: Swaine, 1969.)

**TABLE 1.6   Trace metals in natural soil solutions obtained by
centrifugation from different soils**

Trace Element	Range (μg/L)
Cadmium (Cd)	3.0–5.0
Chromium (Cr)	0.4–0.7
Cobalt (Co)	0.3–5.0
Copper (Cu)	25–140
Lead (Pb)	2–8
Manganese (Mn)	30–270
Molybdenum (Mo)	2–30
Nickel (Ni)	15–150
Zinc (Zn)	20–350

Source: Kabata-Pendias and Pendias, 1994.

**TABLE 1.7   Ranges and medium concentration of trace elements in
dry digested sewage sludges**

	mg/kg		
	Reported Range		
Element	Minimum	Maximum	Median
Arsenic (As)	1.1	230	10
Cadmium (Cd)	1.0	3,400	10
Chromium (Cr)	10	99,000	500
Cobalt (Co)	11.3	2,490	30
Copper (Cu)	84	17,000	800
Fluorine (F)	80	33,500	260
Iron (Fe)	1,000	154,000	17,000
Lead (Pb)	13	26,000	500
Manganese (Mn)	32	9,870	260
Mercury (Hg)	0.6	56	6
Molybdenum (Mo)	0.1	214	4
Nickel (Ni)	2	5,300	80
Selenium (Se)	1.7	17.2	5
Tin (Sn)	2.6	329	14
Zinc (Zn)	101	49,000	1,700

Source: Chaney, 1983.

The recommended upper limits for ten trace elements which can be potentially toxic to plants are given in Table 1.8.

TABLE 1.8 Recommended upper limits of potentially toxic metals in soil plow depth after application of sewage sludge and maximum rate of metal applications

Trace Element	Typical Total Metal Content of Uncontaminated Soil (mg/kg)	Recommended Upper Limit for Soil Concentrations after Sludge Additions (mg/kg)	Maximum Annual Rate of Metal Addition Based on a 10-Year Average (kg/ha/year)
Arsenic (As)	10	20	0.7
Cadmium (Cd)	0.5	3	0.15
Chromium (Cr)	50	600	40
Copper (Cu)	20	135	7.5
Lead (Pb)	50	250	15
Mercury (Hg)	0.1	1	0.1
Molybdenum (Mo)	1	4	0.2
Nickel (Ni)	25	75	3
Selenium (Se)	0.5	3	0.15
Zinc (Zn)	80	300	15

Source: Kabata-Pendias and Pendias, 1994.

A number of countries have specified a maximum acceptable concentrations (MAC) of trace elements in agricultural soils that would be considered phytotoxic to plants (Table 1.9).

In addition, with an established MAC, a maximum application limit (MAL) has also been set by the United Kingdom, the European Community, and Poland (Table 1.10).

Although rate applications of sewage sludge or any other similar trace-element-containing waste product may be based on some set load limit, cumulative trace element (heavy metal) additions can also be set based on the cation exchange capacity (CEC, meq/100 g) of the soil. Load limits increase with increasing soil CEC. Such cumulative heavy metal additions determined by Logan and Chaney (1983) are given in Table 1.11.

Load limits will also be determined by plant species, as the tolerance of plants to various trace elements varies considerably and therefore would be

TABLE 1.9   Proposals for maximum acceptable concentration (MAC) of trace metals considered as phytotoxic in agricultural soils

Trace Metal	*mg/kg*					
	Austria	*Canada*	*Poland*	*Japan*	*United Kingdom*[a]	*Germany*[b]
Arsenic (As)	50	25	30	15	20	40 (50)
Beryllium (Be)	10	—	10	—	—	10 (20)
Cadmium (Cd)	5	8	3	—	1 (3)	2 (5)
Chromium (Cr)	100	75	100	—	20	200 (500)
Cobalt (Co)	50	25	50	50	—	—
Copper (Cu)	100	100	100	125	50 (100)	50 (500)
Lead (Pb)	100	200	100	400	50 (100)	500 (1000)
Mercury (Hg)	5	0.3	5	—	2	10 (50)
Molybdenum (Mo)	10	2	10	—	—	—
Nickel (Ni)	100	100	100	400	30 (50)	100 (200)
Thallium (Tl)	—	—	—	—	—	2 (20)
Zinc (Zn)	300	400	300	250	150 (300)	300 (600)

[a] Values are proposals of the European Economic Community for MAC in soils treated with sewage sludges. Values in parentheses are mandatory concentrations.
[b] Tolerable and toxic (in parentheses) contents.

Source: Kabata-Pendias and Pendias, 1994.

TABLE 1.10   Maximum acceptable concentration (MAC) and maximum application limit (MAL) of trace metals to arable soil

Trace Metal	MAC (mg/kg)	MAL (kg/ha/year)		
		United Kingdom	*European Community*	*Poland*
Zinc (Zn)	2000	15	30	10
Chromium (Cr)	1000	—	—	15
Lead (Pb)	1000	15	15	10
Copper (Cu)	1000	7.5	5	5
Nickel (Ni)	150	3	3	3
Cadmium (Cd)	20	0.15	0.15	0.2
Mercury (Hg)	10	0.1	0.1	0.2

Source: Kabata-Pendias and Pendias, 1994.

TABLE 1.11   Cumulative heavy metal additions in soil based on cation exchange capacity (meq/100 g) of the soil

Heavy Metal	Cumulative Heavy Metal Additions (kg/ha)		
	<5 meq/100 g	*5–15 meq/100 g*	*>15 meq/100 g*
Cadmium (Cd)	5.6	11.2	22.4
Copper (Cu)	140	280	560
Lead (Pb)	560	1120	2240
Nickel (Ni)	56	112	224
Zinc (Zn)	280	560	1120

Source: Logan and Chaney, 1983.

another factor affecting application rates of sewage sludge or other similar trace-element-containing waste products applied to land. A list of some of these plants species is given in Table 1.12.

Forms of the Trace Elements

The form of a trace element as it exists in soil and water can determine its degree of toxicity, as shown in Table 1.13. The best examples of this phenomenon are chromium and mercury, where the most toxic forms of these two trace elements are the 6+ valent chromium and methyl mercury (CH_3Hg). In most other cases, concentration in the environment primarily determines toxicity.

The Oceans

Since approximately two-thirds of the world's surface is covered by oceans, the elemental content of sea water in the major oceans would be of considerable interest, as fish and other substances are gathered from the oceans in considerable quantities. Sea water is also used as a source of some elements and is being evaluated for use as irrigation water under controlled conditions. The elemental content of sea water varies among the major oceans (e.g., Pacific and Atlantic) as well as by depth. The general elemental content of sea water in the oceans is given in Table 1.14.

The trace element content of sea water for the Atlantic and Pacific oceans is given in Table 1.15.

TABLE 1.12 Relative sensitivity of crops to sludge-applied heavy metals[a]

Very Sensitive[b]	Sensitive[c]	Tolerant[d]	Very Tolerant[e]
Chard	Mustard	Cauliflower	Corn
Lettuce	Kale	Cucumber	Sudangrass
Red beet	Spinach	Zucchini squash	Smooth bromegrass
Carrot	Broccoli	Flat pea	'Merlin' red fescue
Turnip	Radish	Oat	
Peanut	Tomato	Orchard grass	
Lando clover	Marigold	Japanese bromegrass	
Alsike clover	Zigzag, red, jura and	Switchgrass	
Crown vetch	crimson clover	Redtop	
'Arc' alfalfa	Alfalfa	Buffalo grass	
White sweetclover	Korean lespedeza	Tall fescue	
Yellow sweetclover	Serica lespedeza	Red fescue	
Weeping lovegrass	Blue lupin	Kentucky bluegrass	
Lehman lovegrass	Birdsfoot trefoil		
Deertongue	Hairy vetch		
Soybean			
Snapbean			
Timothy			
Colonial bentgrass			
Perennial ryegrass			
Creeping bentgrass			

[a] Sassafras sandy loam amended with a highly stabilized and leached digested sludge containing 5300 mg zinc, 2400 mg copper, 320 mg nickel, 390 mg manganese, and 23 mg cadmium per kilogram dry sludge. At 5% sludge, maximum cumulative recommended applications of zinc and copper are made.
[b] Injured at 10% of the high-metal sludge at pH 6.5 and at pH 5.5.
[c] Injured at 10% of the high-metal sludge at pH 5.5 but not at pH 6.5.
[d] Injured at 25% of the high-metal sludge at pH 5.5 and pH 6.5 but not at 10% sludge at pH 5.5 or 6.5.
[e] Not injured at 25% sludge at pH 5.5.

Source: Logan and Chaney, 1983.

Fresh Water

In this section, the trace element content of fresh water from various sources is considered. Much of the data is from the United States. The relative abundance of elements in potable water, including the trace elements, is given in Table 1.16.

TABLE 1.13 Important environmental chemical species of the trace elements

Trace Element	Dominant Chemical Species[a]		Most Toxic Species[b]
	Soil	*Water*	
Antimony (Sb)	$Sb^{III}O_x$?	$Sb(OH)^{6-}$?	?
Arsenic (As)	AsO_4^{3-}	AsO_4^{3-}; AsO_3^{3-}	AsO_4^{3-}
Barium (Ba)	Ba^{2+}	Ba^{2+}	Ba^{2+}
Beryllium (Be)	Be^{2+}; $Be_xO_y^{2x-2y}$	Be^{2+}	Be^{2+}
Bismuth (Bi)	Bi^{3+}?	Bi^{3+}?	?
Boron (B)	$B(OH)_3$	$B(OH)_3$	$B(OH)_3$
Cadmium (Cd)	Cd^{2+}	Cd^{2+}	Cd^{2+}
Chromium (Cr)	Cr^{3+}	Cr^{3+}; Cr^{6+}	Cr^{6+}
Cobalt (Co)	Co^{2+}	Co^{2+}	Co^{2+}
Copper (Cu)	Cu^{2+}	Cu^{2+}-fulvate	Cu^{2+}
Lead (Pb)	Pb^{2+}	$Pb(OH)^+$	Pb^{2+}
Manganese (Mn)	Mn^{4+}; Mn^{2+}	Mn^{2+}	Mn^{2+}
Mercury (Hg)	Hg^{2+}; CH_3Hg	$Hg(OH)_2$·, $HgCl_2$	CH_3Hg
Molybdenum (Mo)	MoO_4^{2-}	MoO_4^{2-}	MoO_4^{2-}
Nickel (Ni)	Ni^{2+}	Ni^{2+}	Ni^{2+}
Selenium (Se)	H_2SeO_3; SeO_4^{2-}	SeO_4^{2-}	SeO_4^{2-}
Silver (Ag)	Ag^+	$Ag+$	Ag^+
Tin (Sn)	$Sn(OH)_6^{2-}$?	$Sn(OH)_6^{2-}$?	?
Tungsten (W)	WO_4^{2-}	WO_4^{2-}?	?
Vanadium (V)	$V^{IV}O_x$?	?	?
Zinc (Zn)	Zn^{2+}	Zn^{2+}	Zn^{2+}

[a] Does not account for ion pairs or complex ion species.
[b] Considers degree of bioavailability.

Source: Allen et al., 1993.

The range in trace element content of groundwater was determined by the U.S. Office of Technology Assessment in 1984 (van der Leeden et al., 1990); the data are given in Table 1.17.

Surface water quality changes are occurring as a result of man's activity. The trace element content changes for the period 1974–81 as reported by the U.S. Geological Survey are shown in Table 1.18.

Markert (1994b) published the elemental content of what he identified as reference fresh water. The trace element content for this water is given in Table 1.19 and the main chemical form (speciation) of the trace elements is given in Table 1.20.

TABLE 1.14 The abundance of the elements in sea water in the oceans

Level	Elements
$>10^6$ mM	Calcium (Ca)
	Carbon (C)
	Chlorine (Cl)
	Hydrogen (H)
	Magnesium (Mg)
	Nitrogen (N)
	Oxygen (O)
	Potassium (K)
	Sodium (Na)
	Sulfur (S)
10^6–10^2 mM	Barium (Ba)
	Boron (B)
	Bromine (Br)
	Fluorine (F)
	Iodine (I)
	Lithium (Li)
	Phosphorus (P)
	Rubidium (Rb)
	Silicon (Si)
	Strontium (Sr)
10^2–1 mM	Aluminum (Al)
	Antimony (Sb)
	Cadmium (Cd)
	Cesium (Cs)
	Chromium (Cr)
	Cobalt (Co)
	Copper (Cu)
	Iron (Fe)
	Manganese (Mn)
	Molybdenum (Mo)
	Nickel (Ni)
	Selenium (Se)
	Titanium (Ti)
	Tungsten (W)
	Uranium (U)
	Vanadium (V)
	Zinc (Zn)

TABLE 1.15 Trace element content in sea water

| | Atlantic | | Pacific | | |
Trace Element	Surface	Deep	Surface	Deep	Other
			mg/L		
Aluminum (Al)	9.7×10^{-4}	5.2×10^{-4}	1.3×10^{-4}	0.13×10^{-4}	—
Antimony (Sb)	—	—	—	—	0.3×10^{-3}
Arsenic (As)	1.45×10^{-3}	1.53×10^{-3}	1.45×10^{-3}	1.75×10^{-3}	—
Barium (Ba)	4.7×10^{-3}	9.3×10^{-3}	4.7×10^{-3}	20.0×10^{-3}	—
Beryllium (Be)	8.8×10^{-8}	17.5×10^{-8}	3.5×10^{-8}	22.0×10^{-8}	—
Bismuth (Bi)	5.1×10^{-8}	—	4.0×10^{-8}	0.4×10^{-8}	—
Boron (B)	—	—	—	—	4.41
Cadmium (Cd)	1.1×10^{-6}	38.0×10^{-6}	1.1×10^{-6}	100.0×10^{-6}	—
Cesium (Cs)	—	—	—	—	3.0×10^{-4}
Chromium (Cr)	1.8×10^{-4}	2.3×10^{-4}	1.5×10^{-4}	2.5×10^{-4}	—
Cobalt (Co)	—	—	6.9×10^{-6}	1.1×10^{-6}	—
Copper (Cu)	8.0×10^{-5}	12.0×10^{-5}	8.0×10^{-5}	28.0×10^{-5}	—
Fluorine (F)	—	—	—	—	1.3
Gallium (Ga)	—	—	—	—	3.0×10^{-5}
Germanium (Ge)	0.07×10^{-6}	0.14×10^{-6}	0.35×10^{-6}	7.0×10^{-6}	—
Gold (Au)	—	—	—	—	1.0×10^{-5}
Hafnium (Hf)	—	—	—	—	7.0×10^{-6}
Indium (In)	—	—	—	—	1.0×10^{-7}
Iodine (I)	0.049	0.0056	0.043	0.058	—
Iron (Fe)	1.0×10^{-4}	4.0×10^{-4}	0.1×10^{-4}	1.0×10^{-4}	—
Lead (Pb)	30.0×10^{-6}	4.0×10^{-6}	10.0×10^{-6}	1.0×10^{-6}	—
Lithium (Li)	—	—	—	—	0.17

TABLE 1.15 Trace element content in sea water (continued)

| | Atlantic | | Pacific | | |
Trace Element	Surface	Deep	Surface	Deep	Other
Manganese (Mn)	1.0×10^{-4}	0.96×10^{-4}	1.0×10^{-4}	0.4×10^{-4}	—
Mercury (Hg)	4.9×10^{-7}	4.9×10^{-7}	3.3×10^{-7}	3.3×10^{-7}	—
Molybdenum (Mo)	—	—	—	—	0.01
Nickel (Ni)	1.0×10^{-4}	4.0×10^{-4}	1.0×10^{-4}	5.7×10^{-4}	—
Platinum (Pt)	—	—	1.1×10^{-7}	2.7×10^{-7}	—
Rubidium (Rb)	—	—	—	—	0.12
Selenium (Se)	0.46×10^{-7}	1.8×10^{-7}	0.15×10^{-7}	1.65×10^{-7}	—
Silicon (Si)	0.03	0.82	0.03	4.09	—
Silver (Ag)	—	—	1.0×10^{-7}	24.0×10^{-7}	—
Strontium (Sr)	7.6	7.7	7.6	7.7	—
Thallium (Tl)	—	—	—	—	1.4×10^{-5}
Tin (Sn)	2.3×10^{-6}	5.8×10^{-6}	—	—	—
Titanium (Ti)	—	—	—	—	4.8×10^{-4}
Uranium (U)	—	—	—	—	3.13×10^{-3}
Vanadium (V)	1.1×10^{-3}	—	1.6×10^{-3}	1.8×10^{-3}	—
Zinc (Zn)	0.05×10^{-4}	1.0×10^{-4}	0.5×10^{-4}	5.2×10^{-4}	—

mg/L

Source: Emsley, 1991.

TABLE 1.16 Relative abundance of elements in potable water, from classification based upon relative abundance of the elements

Element Contents (mg/L)			
Major (1.0–1000)	*Secondary* (0.01–10.0)	*Minor* (0.0001–0.1)	*Trace* (<0.001)
Sodium	Iron	Antimony[a]	Beryllium
Calcium	Strontium	Aluminum	Bismuth
Magnesium	Potassium	Arsenic	Cerium[a]
Bicarbonate	Carbonate	Barium	Cesium
Sulfate	Nitrate	Bromide	Gallium
Chloride	Fluoride	Cadmium[a]	Gold
	Silicon	Boron	Chromium[a]
	Indium	Cobalt	Lanthanum
		Copper	Niobium[a]
		Germanium[a]	Platinum
		Iodide	Radium
		Lead	Ruthenium[a]
		Lithium	Scandium[a]
			Manganese
			Silver
			Molybdenum
			Thallium[a]
			Nickel
			Thorium[a]
			Phosphate
			Tin
			Rubidium[a]
			Tungsten[a]
			Selenium
			Ytterbium
			Titanium[a]
			Yttrium[a]
			Uranium
			Zirconium[a]
			Vanadium
			Zinc

[a] These elements occupy an uncertain position in the list.

Source: Davis and DeWiest, 1966.

TABLE 1.17   Trace elements found in groundwater

Trace Element	Concentration (mg/L)	Trace Element	Concentration (mg/L)
Aluminum (Al)	0.1–1200	Iron (Fe)	0.04–6200
Arsenic (As)	0.01–2100	Lead (Pb)	0.01–5.6
Barium (Ba)	2.8–3.8	Manganese (Mn)	0.1–110
Beryllium (Be)	<0.01	Mercury (Hg)	0.003–0.01
Cadmium (Cd)	0.01–180	Nickel (Ni)	0.05–0.5
Chromium (Cr)	0.06–2740	Selenium (Se)	0.6–20
Cobalt (Co)	0.01–0.18	Silver (Ag)	9.0–330
Copper (Cu)	0.01–2.8	Vanadium (V)	0.1–243
Fluoride (F)	0.1–250	Zinc (Zn)	0.1–240

TABLE 1.18   Trends of surface water quality in the United States 1974–81 (selected water quality trace elements at NASCAN stations)

Trace Element	Number of Stations with			
	Increasing Trends	No Change	Decreasing Trends	Total Stations
Arsenic (As)	68	228	11	307
Barium (Ba)	4	81	1	86
Boron (B)	2	15	3	20
Cadmium (Cd)	32	264	7	303
Chloride (Cl)	104	164	36	304
Chromium (Cr)	12	152	2	166
Copper (Cu)	6	83	6	95
Iron (Fe)	28	258	21	307
Lead (Pb)	5	232	76	313
Manganese (Mn)	30	250	19	299
Mercury (Hg)	8	194	2	204
Selenium (Se)	2	201	21	224
Silicon (Si)	48	213	41	302
Silver (Ag)	1	32	0	33
Zinc (Zn)	19	251	32	302

Source: U.S. Geological Survey Water-Supply Paper 2250, 1981.

TABLE 1.19 Trace element content of "reference fresh water"

Trace Element	µg/L	Trace Element	µg/L
Enzymatic Element of Transition Metals		*Main Group—4th*	
Cobalt (Co)	0.5	Tin (Sn)	0.01
Chromium (Cr)	1.0	Lead (Pb)	3.0
Copper (Cu)	3.0		
Iron (Fe)	500	*Main Group—5th*	
Manganese (Mn)	5.0	Arsenic (As)	0.5
Molybdenum (Mo)	1.0	Antimony (Sb)	0.2
Nickel (Ni)	0.3	Bismuth (Bi)	0.05
Vanadium (V)	1.0		
Zinc (Zn)	5.0	*Main Group—6th*	
		Selenium (Se)	0.2
Main Group—1st			
Rubidium (Rb)	1.0	*Main Group—7th*	
Cesium (Cs)	0.005	Bromine (Br)	15
		Iodine (I)	3.0
Main Group—2nd			
Beryllium (Be)	0.1	*Subgroup—1st*	
Strontium (Sr)	50	Silver (Ag)	0.3
Barium (Ba)	10	Gold (Au)	0.002
Main Group—3rd		*Subgroup—2nd*	
Boron (B)	10	Cadmium (Cd)	0.2
Aluminum (Al)	200	Mercury (Hg)	0.1
Gallium (Ga)	0.1		
Thallium (Tl)	0.04		

Note: No data for highly polluted water were used. The sequence of the elements is based on their position in the periodic table. Exceptions are the transitional metals (cobalt, chromium, copper, iron, manganese, nickel, and zinc), which have an enzymatic effect.

Source: Markert and Geller, 1994.

Drinking Water

The quality of drinking water is a major ecological problem, as many parts of the world lack clean, reliable drinking water supplies. Acceptable levels for the trace elements have been established by various agencies. Those levels established as the National Primary Drinking Water Standards set by the U.S. Environmental Protection Agency (EPA) are given in Table 1.21, and the Secondary Standards are listed in Table 1.22.

TABLE 1.20 Mainly observed forms of trace elements in fresh water (speciation)

Trace Element	Main Forms of Chemical Species
Aluminum (Al)	$Al(OH)_4^-$, possibly: Al^{3+}, $AlOH^{2+}$, $Al(OH)^{2+}$, $Al(OH)_3$
Antimony (Sb)	$Sb(OH)_6^-$
Arsenic (As)	$HAsO_4^{2-}$, $H_2AsO_4^-$, possibly: H_3AsO_4, AsO_4^{3-}, or $H_2AsO_3^-$
Barium (Ba)	Ba^{2+}
Beryllium (Be)	$Be(OH)_2^+$
Boron (B)	$B(OH)_3$ or $B(OH)_4^-$
Cadmium (Cd)	Cd^{2+} and $CdOH^+$
Cerium (Ce)	Ce^{3+} or $CeOH^{2+}$
Cesium (Cs)	Cs^+
Chromium (Cr)	CrO_4^{2-} or $Cr(OH)_3$
Cobalt (Co)	Co^{2+} and $CoCO_3$
Copper (Cu)	$CuOH^+$ or $CuCO_3$
Gallium (Ga)	$Ga(OH)_4^-$
Germanium (Ge)	$Ge(OH)_4$
Gold (Au)	$Au(OH)_4^-$
Iron (Fe)	$Fe(OH)_2^+$ in oxygen-containing areas Fe^{2+} under reducing conditions
Lead (Pb)	$PbCO_3$ or as $Pb(CO_3)_2^{2-}$
Lithium (Li)	Li^+
Manganese (Mn)	Mn^{2+}
Mercury (Hg)	$Hg(OH)_2$ and $HgOHCl$
Molybdenum (Mo)	MoO_4^{2-}
Nickel (Ni)	Ni^{2+}, also $NiCO_3$
Rubidium (Rb)	Rb^+
Selenium (Se)	SeO_3^{2-}, possibly: $HSeO_3^-$, H_2SeO_3, SeO_4^{2-}, and $HSeO_4^-$
Silver (Ag)	Ag^+
Strontium (Sr)	Sr^{2+}, possibly $SrOH^+$
Thallium (Tl)	Tl^+
Tin (Sn)	Mono-, di-, and trimethyl compounds
Titanium (Ti)	$Ti(OH)_4$
Vanadium (V)	$H_2VO_4^-$ or HVO_4^{2-}
Zinc (Zn)	$ZnOH^+$, Zn^{2+}, or $ZnCO_3$

Note: No species linked to colloids or humic substances were considered.

Source: Markert and Geller, 1994.

TABLE 1.21 National Primary Drinking Water Standards

Trace Element	MCL Level[a] (mg/L)
Arsenic (As)	0.05
Barium (Ba)	1.0
Cadmium (Cd)	0.01
Chromium (Cr)	0.05
Fluoride (F)	4.0
Lead (Pb)	0.05
Mercury (Hg)	0.002
Selenium (Se)	0.01
Silver (Ag)	0.05

[a] MCL = maximum contaminant level.

Source: U.S. Environmental Protection Agency.

TABLE 1.22 National Secondary Drinking Water Standards

Trace Element	SMCL Level[a] (mg/L)
Chloride (Cl)	250
Copper (Cu)	1.0
Fluoride (F)	2.0
Iron (Fe)	0.3
Manganese (Mn)	0.05
Zinc (Zn)	5.0

[a] SMCL = secondary maximum contaminant level.

Source: U.S. Environmental Protection Agency.

TABLE 1.23 Canadian guidelines for maximum acceptable concentration of trace elements in drinking water

Trace Element	Concentration (mg/L)
Arsenic (As)	0.05
Barium (Ba)	1.0
Boron (B)	5.0
Cadmium (Cd)	0.005
Chloride (Cl)	250
Chromium (Cr)	0.05
Copper (Cu)	1.0
Fluoride (F)	1.5
Iron (Fe)	0.3
Lead (Pb)	0.05
Manganese (Mn)	0.05
Mercury (Hg)	0.001
Selenium (Se)	0.01
Silver (Ag)	0.05
Uranium (U)	0.02
Zinc (Zn)	5.0

TABLE 1.24   WHO guidelines for drinking water quality

Trace Element	Guideline Value (mg/L)
Aluminum (Al)	0.2
Chloride (Cl)	250
Copper (Cu)	1.0
Iron (Fe)	0.3
Manganese (Mn)	0.1
Zinc (Zn)	5.0

The maximum acceptable concentrations of trace elements in drinking water set by Canada are listed in Table 1.23, and the guideline values for trace elements set by the World Health Organization (WHO) are given in Table 1.24.

A comparison of the primary drinking water regulations among the United States, Canada, the European Economic Community (EEC), and the WHO is given in Table 1.25.

TABLE 1.25   Comparison of U.S. primary drinking water regulations with Canadian, EEC, and WHO guidelines

	mg/L		
Trace Element	U.S. Maximum Contaminant Level	Canadian Maximum Acceptable Limit	EEC Maximum Admissible Concentration	WHO Guideline Value
Arsenic (As)	0.05	0.05	0.05	0.05
Barium (Ba)	1.0	1.0	0.1	NG
Cadmium (Cd)	0.01	0.005	0.005	0.005
Chromium (Cr)	0.05	0.05	0.05	0.05
Fluoride (F)	4.0	1.5	NG	1.5
Lead (Pb)	0.05	0.05	0.05	0.05
Mercury (Hg)	0.002	0.001	0.001	0.001
Selenium (Se)	0.01	0.01	0.01	0.01
Silver (Ag)	0.05	0.05	0.01	NG

NG = not given.

A similar comparison among the drinking water maximum contaminant levels of trace elements for the United States, EEC, and Australia is given in Table 1.26.

TABLE 1.26 Maximum contaminant levels for trace elements in drinking water set by the U.S. EPA, the EEC, and the Australian Water Resources Council

	μg/L		
Trace Element	*U.S. EPA[a]*	*EEC[b]*	*Guideline Values[c]*
Aluminum (Al)	—	200	200
Antimony (Sb)	—	10	—
Arsenic (As)	—	200	50
Barium (Ba)	1000	1000	—
Beryllium (Be)	1	—	—
Cadmium (Cd)	10	5	5
Chromium (Cr)	50	50	50
Copper (Cu)	1000	—	1000
Iron (Fe)	300	200	300
Lead (Pb)	50	50	50
Manganese (Mn)	100	50	100
Nickel (Ni)	50	50	50
Selenium (Se)	10	10	—
Silver (Ag)	50	10	—
Thallium (Tl)	2	—	—
Zinc (Zn)	5000	5000	5000

[a] U.S. EPA, *Federal Register* 40 (141), National Primary Drinking Water Regulations, 1975; *Federal Register* 44 (140), National Secondary Drinking Water Regulations, 1979.
[b] Official Journal European Communities No. L229, pp. 11–29, August 25, 1990.
[c] Guidelines for drinking water quality in Australia. National Health and Medical Research Council and Australian Water Resources Council, 1987.

In 1962, a quality survey was conducted to evaluate public water supplies for 100 of the largest cities in the United States. The trace element content results of that survey are given in Table 1.27, and the range in trace element content is given in Table 1.28.

In 1984, the EPA determined the trace element content of rural water supplies; the results are given in Table 1.29.

TABLE 1.27   Quantity of trace element limits of finished water in public water supplies of the 100 largest cities in the United States

Trace Element	Water Supplies Having Less than Stated Concentration	
	Concentration (mg/L)	Percent of Water Supplies
Chloride (Cl)	50.0	93
Fluoride (F)	1.0	92
Iron (Fe)	0.25	98
Manganese (Mn)	0.10	95
Silica (SiO$_2$)	30.0	94
	Concentration (μg/L)	Percent of Water Supplies
Aluminum (Al)	500	87
Barium (Ba)	100	94
Boron (B)	100	94
Chromium (Cr)	5	95
Copper (Cu)	100	94
Iron (Fe)	150	94
Lead (Pb)	10	95
Lithium (Li)	50	96
Manganese (Mn)	100	97
Molybdenum (Mo)	10	96
Nickel (Ni)	10	95
Rubidium (Rb)	5	91
Silica (SiO$_2$)	30	94
Silver (Ag)	0.5	95
Strontium (Sr)	500	96
Titanium (Ti)	5	96
Vanadium (V)	10	91

Note: Data as of 1962.

The trace element content of drinking water in three Hungarian cities is given in Table 1.30.

Irrigation Water

The quality of irrigation water can determine its usefulness. Therefore, the Food and Agricultural Organization (FAO) of the United Nations has estab-

TABLE 1.28   Range in trace element content of finished water in public water supplies of the 100 largest cities in the United States as of 1962

	mg/L		
Trace Element	*Maximum*	*Median*	*Minimum*
Chloride (Cl)	540	13.0	0.0
Fluoride (F)	7.0	0.4	0.0
Iron (Fe)	1.30	0.02	0.00
Manganese (Mn)	2.50	0.00	0.00
Silica (SiO_2)	72	7.1	0.0
	μg/L		
Aluminum (Al)	1500	54	3.3
Barium (Ba)	380	43	1.7
Boron (B)	590	31	2.5
Chromium (Cr)	35	0.43	ND
Copper (Cu)	250	8.3	<0.61
Iron (Fe)	1700	43	1.9
Lead (Pb)	62	3.7	ND
Lithium (Li)	170	2.0	ND
Manganese (Mn)	1100	5.0	ND
Molybdenum (Mo)	68	1.4	ND
Nickel (Ni)	34	<2.7	ND
Rubidium (Rb)	67	1.05	ND
Silver (Ag)	7.0	0.23	ND
Strontium (Sr)	1200	110	2.2
Titanium (Ti)	49	<1.5	ND
Vanadium (V)	70	<4.3	ND

ND = not detected.

lished the maximum concentration of trace elements in irrigation water as shown in Table 1.31.

Livestock Water Quality

The suitability of water for livestock has been established, and the trace element maximums are given in Table 1.32.

TABLE 1.29 Summary of trace elements found in rural water supplies (according to survey conducted by U.S. EPA)

Trace Element	Level Exceeded (mg/L)	In Percent of Rural Households				
		Nationwide	West	North-Central	Northeast	South
Arsenic (As)	0.05	0.8	2.1	1.8	0.0	0.0
Barium (Ba)	1.0	0.3	0.0	0.0	0.0	0.7
Cadmium (Cd)	0.01	16.8	27.1	20.7	1.6	17.3
Chromium (Cr)	0.05	[a]	0.0	0.0	0.0	0.0
Fluoride (F)	1.4	2.5	6.2	1.8	0.0	2.7
Iron (Fe)	0.3	18.7	7.0	28.2	16.0	17.0
Lead (Pb)	0.05	16.6	16.9[b]	10.8[b]	9.6[b]	23.1[b]
Manganese (Mn)	0.05	14.2	4.7	19.9	16.9	12.3
Mercury (Hg)	0.002	24.1	10.4	31.8	22.0	25.0
Selenium (Se)	0.01	13.7	41.3	25.7	0.0	2.1
Silver (Ag)	0.05	4.7	2.1	3.7	4.8	4.8

[a] Not detected.
[b] May be distorted upward.

Source: U.S. EPA. 1984. National Statistical Assessment of Rural Conditions, Executive Summary. Office of Drinking Water.

TABLE 1.30 Trace element content of drinking water from different Hungarian waterworks

Trace Element	mg/mL		
	Budapest	Szolnok	Gyula
Aluminum (Al)	0.10	0.14	ND
Barium (Ba)	0.15	0.10	0.05
Boron (B)	0.14	0.25	0.10
Copper (Cu)	0.02	0.02	0.01
Iron (Fe)	2.50	0.60	0.20
Manganese (Mn)	0.20	0.16	0.10
Molybdenum (Mo)	ND	0.01	ND
Nickel (Ni)	0.18	0.02	ND
Silicon (Si)	9.20	8.90	10.70
Strontium (Sr)	0.50	0.32	0.24
Zinc (Zn)	2.05	0.12	0.02

ND = not detected.

Source: University of Horticulture and Food Science, Budapest, Hungary.

TABLE 1.31   FAO recommended maximum concentration of trace elements in irrigation water

Trace Element	*Maximum Concentration (mg/L)*
Aluminum (Al)	5.0
Arsenic (As)	0.10
Beryllium (Be)	0.10
Cadmium (Cd)	0.01
Chromium (Cr)	0.10
Cobalt	0.05
Copper (Cu)	0.20
Fluorine (F)	1.0
Iron (Fe)	5.0
Lead (Pb)	5.0
Lithium (Li)	2.5
Manganese (Mn)	0.20
Molybdenum (Mo)	0.01
Nickel (Ni)	0.20
Selenium (Se)	0.02
Vanadium (V)	0.10
Zinc (Zn)	2.0

TABLE 1.32   Water quality criteria for livestock

Trace Element	*Limiting Threshold (mg/L)*
Aluminum (Al)	5.0
Arsenic (As)	0.20
Boron (B)	5.0
Cadmium (Cd)	0.05
Chromium (Cr)	1.0
Cobalt (Co)	1.0
Copper (Cu)	0.50
Fluorine (F)	2.0
Lead (Pb)	1.0
Mercury (Hg)	0.01
Nickel (Ni)	1.0
Selenium (Se)	0.05
Vanadium (V)	0.10
Zinc (Zn)	25.0

BIOLOGICAL CLASSIFICATION AND FUNCTION

Various classification systems have been proposed for those elements—both major as well as the trace elements—considered essential for the normal development and growth of biologicals. A classification system suggested by Frieden (1981), which is illustrated in Table 1.33, divides the elements (macroelements, trace, and ultratrace) into various categories based on the amount found in tissue rather than designating a specific function for essentiality.

Based on current acceptance by the scientific community, a classification system of essentiality for the trace elements/micronutrients considered in this text and identified in periodic table format as shown in Figure 1.1 can be given as follows:

1. *"Classically essential" trace elements*—boron, cobalt, copper, iodine, iron, manganese, molybdenum, and zinc
2. *Probable essential trace elements*—chromium, fluorine, nickel, selenium, and vanadium
3. *Physiologically promotive trace elements*—bromine, lithium, silicon, tin, and titanium

TABLE 1.33   Classification of the essential elements

Classification	*Elements*
Bulk structural elements	Carbon (C), hydrogen (H), oxygen (O), phosphorus (P), and sulfur (S)
Macroelements	Calcium (Ca), chlorine (Cl), potassium (K), and sodium (Na)
Trace elements	Copper (Cu), iron (Fe), and zinc (Zn)
Ultratrace elements	
Non-metals	Arsenic (As), boron (B), fluorine (F), iodine (I), and selenium (Se)
Metals	Cadmium (Cd), chromium (Cr), cobalt (Co), lead (Pb), manganese (Mn), molybdenum (Mo), nickel (Ni), tin (Sn), vanadium (V)

Source: Frieden, 1985.

4. *Elements that have a promotive role and have been partly verified*—aluminum, arsenic, cadmium, cesium, gallium, germanium, lead, platinum, rubidium, strontium, and tungsten
5. *Trace elements that may have a promotive role (under special circumstances)*—antimony, barium, beryllium, bismuth, gold, hafnium, indium, iridium, mercury, niobium, osmium, palladium, rhenium, rhodium, ruthenium, scandium, silver, tantalum, tellurium, thallium, yttrium, zirconium, and the rare earths

The decisions within the scientific community as to which of the many trace elements were essential and/or beneficial did not come into full fruition until the late 1960s and early 1970s, when the analytical procedures for their determination were greatly improved (see Chapter 6). During this period, several additional trace elements were generally accepted as probably essential: selenium, chromium, arsenic, and tin. Discussing the future need for trace element research, Schwarz (1970) stated that "to demonstrate that an element has essentiality will be a very hard task, but to exclude categorically from being essential, has no real basis." Of the 88 permanent elements found in the earth's crust, only 12 are unlikely to be essential to life—the noble gases and the strong radioactive elements.

Mertz (1980) wrote that "in the recent history of nutrition the first half of our century was characterised as the vitamin era and the second part of this century will be known as the era of trace elements" and later defined trace element requirements and current recommendations (Mertz, 1989). Since 1970, 11 additional trace elements have been proposed to be required by animals and possibly humans: arsenic, boron, bromine, cadmium, fluorine, lead, lithium, nickel, silicon, tin, and vanadium. Except for boron and nickel, none of these elements is considered to be essential for plants, although it has been recently suggested that nickel should be classed as essential for higher plants (Eskew et al., 1983; Brown et al., 1987).

Essentiality for a trace element or micronutrient is based on some known or suspected function in an organism, a list of which has been prepared by Kabata-Pendias and Pendias (1984) for a number of elements, as shown in Table 1.34.

A similar identification of the forms and principal function of 19 trace elements in plants has been given by Kabata-Pendias and Pendias (1994) and is shown in Table 1.35. This table might be taken to suggest that there are 19 essential micronutrients rather than the currently accepted 7.

In some of the current literature, such as that by Schmidt (1989) and Gibson (1990), there have been attempts to classify the trace elements into

TABLE 1.34   Functions and forms of the elements in organisms

Function/Form	Elements
Incorporated into structural materials	Silicon, iron, and rarely barium and strontium
Bound into miscellaneous small molecules including antibiotics and porphyrins	Arsenic, boron, bromine, cobalt, copper, fluorine, iodine, iron, mercury, selenium, and vanadium
Combines with large molecules, mainly proteins, including enzymes with catalytic properties	Cobalt, chromium(?), copper, iron, manganese, molybdenum, nickel(?), selenium, and zinc
Fixed by large molecules having storage, transport, or unknown functions	Cadmium, cobalt, copper, iodine, iron, manganese, mercury, nickel, selenium, and zinc
Related to organelles or their parts (e.g., mitochondria, chloroplasts, some enzyme systems)	Copper, iron, manganese, molybdenum, and zinc

Source: Kabata-Pendias and Pendias, 1984.

TABLE 1.35   Forms and principal functions of trace elements that are essential for plants

Element	Constituent	Involved
Aluminum (Al)[a]	—	Controlling colloidal properties in the cell, possible activation of some dehydrogenases and oxidases
Arsenic (As)[a]	Phospholipid (in algae)	Metabolism of carbohydrates in algae and fungi
Boron (B)	Phosphogluconates	Metabolism and transport of carbohydrates, flavonoid synthesis, nucleic acid synthesis, phosphate utilization, and polyphenol production
Bromine (Br)[a]	Bromophenols (in algae)	—
Cobalt (Co)	Cobalamide coenzyme	Symbiotic N_2 fixation, possibly also in non-nodulating plants, and valence changes stimulation synthesis of chlorophyll and proteins (?)

TABLE 1.35   Forms and principal functions of trace elements that are essential for plants (continued)

Element	Constituent	Involved
Copper (Cu)	Various oxidases, plastocarbohydrate	Oxidation, photosynthesis, protein and cyanins, and centroplasmin metabolism, possibly involved in symbiotic N_2 fixation and valence changes
Fluorine (F)	Fluoroacetate (in a few species)	Citrate conversions
Iron (Fe)	Hemoproteins and non-heme	Photosynthesis, N_2 fixation, and valence changes iron proteins, dehydrogenases, and ferredoxins
Lithium (Li)[a]	—	Metabolism in halophytes
Manganese (Mn)	Many enzyme systems	Photoproduction of oxygen in chloroplasts and indirectly on NO_3 reduction
Molybdenum (Mo)	Nitrate reductase, nitrogenase, oxidases, and molybdo-ferredoxin	N_2 fixation, NO_3 reduction, and valence changes
Nickel (Ni)[a]	Enzyme urease (in *Canavalia* seeds)	Possibly in action of hydrogenase and translocation of N
Rubidium (Rb)[a]	—	Function similar to that of K in some plants
Selenium (Se)[a]	Glycene reductase in *Clostridium* cells)	—
Silicon (Si)	Structural components	—
Strontium (Sr)[a]	—	Function similar to that of Ca in some plants
Titanium (Ti)[a]	—	Possibly photosynthesis and N_2 fixation
Vanadium (V)	Porphyrins	Lipid metabolism, photosynthesis (in green algae), and possibly in N_2 fixation
Zinc (Zn)	Anhydrases, dehydrogenases, proteinases, and peptidases	Carbohydrate and protein metabolism

[a] Elements known to be essential for some groups or species and whose general essentiality needs confirmation.

Source: Kabata-Pendias and Pendias, 1994.

TABLE 1.36   Generally and partially accepted essential trace elements

Generally Accepted	*Partially Accepted*
Boron (B)	Arsenic (As)
Cobalt (Co)	Chromium (Cr)
Copper (Cu)	Fluorine (F)
Iodine (I)	Lead (Pb)
Iron (Fe)	Lithium (Li)
Manganese (Mn)	Selenium (Se)
Molybdenum (Mo)	Silicon (Si)
Nickel (Ni)	Tin (Sn)
Zinc (Zn)	Vanadium (V)

Source: Pais, 1989.

two categories: those "generally accepted" as essential and those "partially accepted." For example, as shown in Table 1.36, one can divide 18 trace elements into two groups: those "generally accepted" (boron, cobalt, copper, iodine, iron, manganese, molybdenum, nickel, and zinc) and those in the "partially accepted" group (arsenic, chromium, fluorine, lead, lithium, selenium, silicon, tin, and vanadium). However, it should be understood that this classification of the trace elements is somewhat uncertain, as well as dependent on personal opinion. For example, Pais (1992) does not share the opinion that there are only two groupings for the trace elements: those that are essential and those that are not essential. There may be another category that needs to be established—trace elements that have beneficial effects on the growth and development of organisms (Nielsen, 1984; Nieboer and Stanford, 1985).

To some, it may seem surprising that both arsenic and lead are included in the "partially essential" family of trace elements, although according to widely accepted opinion these elements and their compounds are considered highly toxic under commonly occurring circumstances. In spite of this, their essentiality status is valid, but only under certain conditions and within very low and narrow concentration limits.

Adriano (1986) has also classified a number of the trace elements into those that are either essential, beneficial, or toxic to plants and animals, as shown in Table 1.37.

As stated earlier, most of the trace elements (i.e., boron, copper, manganese, and zinc) identified as essential for plants can also be toxic when above certain concentration levels. In addition, many of the other trace

TABLE 1.37   Classification of trace metals as plant and animal nutrients or toxins

Trace Element	Essential to Plants	Beneficial to Animals	Toxic to Plants	Toxic to Animals
Antimony (Sb)	No	No	?	Yes
Arsenic (As)	No	Yes	Yes	Yes
Barium (Ba)	No	Possible	Low	Low
Beryllium (Be)	No	No	Yes	Yes
Bismuth (Bi)	No	No	Yes	Yes
Boron (B)	Yes	No	Yes	—
Cadmium (Cd)	No	No	Yes	Yes
Chromium (Cr)	No	Yes	Yes	Yes (Cr^{6+})
Cobalt (Co)	Yes	Yes	Low	Low
Copper (Cu)	Yes	Yes	Yes	Yes
Lead (Pb)	No	No	Yes	Yes
Manganese (Mn)	Yes	Yes	Yes	Low
Mercury (Hg)	No	No	No	Yes
Molybdenum (Mo)	Yes	Yes	Yes	Yes
Nickel (Ni)	Possible	Yes	Yes	Yes
Selenium (Se)	Yes	Yes	Yes	Yes
Silver (Ag)	No	No	No	Yes
Tin (Sn)	No	Yes	?	Yes
Tungsten (W)	No	No	?	?
Vanadium (V)	Yes	Yes	Yes	Yes
Zinc (Zn)	Yes	Yes	Yes	Yes

Note: Toxicity considers the likelihood of uptake of the metal.

Based in part on Adriano, 1986.

elements not identified as essential can also be toxic for plants, animals, and humans when above certain levels. Biological monitoring of toxic trace elements has been evaluated by Jenkins (1981). Those identified as being the most toxic are mercury, copper, cobalt, cadmium, boron, possibly silver, and tin. The toxic effects of the various trace elements on an organism are summarized in Table 1.38.

PLANT PHYSIOLOGY

Based on the rigid requirements established by Arnon and Stout (1939) for defining essentiality for plants, since 1940 the requirements have been:

TABLE 1.38   Basic reactions related to toxic effects of excess trace element

Trace Element	Reaction
Bromine (Br), cadmium (Cd), copper (Cu), fluorine (F), gold (Au), iodine (I), lead (Pb), mercury (Hg), silver (Ag)	Changes in permeability of the cell membrane
Lead (Pb), mercury (Hg), silver (Ag)	Reactions of thiol groups with cations
Arsenic (As), antimony (Sb), fluorine (F), selenium (Se), tellurium (Te), tungsten (W)	Competition for sites with essential metabolites
Aluminum (Al), beryllium (Be), scandium (Sc), yttrium (Y), zirconium (Zr)	Affinity for reaction with phosphate groups and active groups of ADP or ATP
Cesium (Cs), lithium (Li), rubidium (Rb), selenium (Se), strontium (Sr)	Replacement of essential ions (mainly major cations)
Arsenate, borate, bromate, fluorate, selenate, tellurate, and tungstate	Occupation of sites for essential groups such as phosphate and nitrate

1. The organism can neither grow nor complete its life cycle in the absence of the element.
2. The element cannot be replaced completely by any other element.
3. The given element has direct influence on the organism and is involved in its metabolism.

Plant physiologists have accepted only seven trace elements (which have since been called micronutrients) as essential: boron, chlorine, copper, iron, manganese, molybdenum, and zinc. However, all the micronutrients except chlorine had been identified as essential for plants before the Arnon and Stout (1939) criteria were established, which may have capped additions to the list. The discoverer of essentiality and the year of discovery are given in Table 1.39.

It has been recently suggested that nickel should be classed as essential for higher plants (Eskew et al., 1983; Brown et al., 1987) and therefore should be added to the current list of seven. The trace elements listed in Table 1.36 would suggest that there are 18 essential micronutrients rather than the generally accepted 7, although all the criteria of Arnon and Stout (1939) have yet to be positively proven for the 12 (arsenic, chromium, cobalt, fluorine, iodine, lead, lithium, nickel, selenium, silicon, tin, and vanadium).

TABLE 1.39   Discoverer of essentiality and year for the micronutrients

Micronutrient	Discoverer of Essentiality	Year
Iron (Fe)	van Sacks, Knop	1860
Manganese (Mn)	McHargue	1922
Zinc (Zn)	Sommer and Lipman	1926
Copper (Cu)	Sommer	1931
	Lipman and MacKinnon	1931
Boron (B)	Sommer and Lipman	1926
Molybdenum (Mo)	Arnon and Stout	1939
Chlorine (Cl)	Stout	1954

Source: Glass, 1989.

There are also several trace elements that have been found to have beneficial effects on plants: silicon, which affects plant stalk strength; cobalt, for nitrogen (N_2) fixation in legumes; and the substitution by vanadium that occurs for molybdenum. A discussion of these elements and their beneficial effects has been presented in a review article by Asher (1991).

For the hydroponic culture of plants, those who initially used the procedure for either (1) studying plant nutrition or (2) the commercial production of plants included a suite of trace elements in two solutions, A and B, which were identified as the A–Z Solutions (Hewitt, 1966). Their compositions are given in Table 1.40. Evidently these early researchers recognized that there were many elements that could beneficially affect plant growth and development, and hence their inclusion in the nutrient solution.

The relative concentration relationship among all the essential elements (major elements and micronutrients) in a plant has been given by Epstein (1965), as shown in Table 1.41.

Porter and Lawlor (1991) also assembled information on the micronutrients, relating plant content (from Epstein, 1965) with that in the environment plus function in the plant, as shown in Table 1.42.

There is much data in the literature on the trace element content of plants associated with different growing and site conditions. Much of this data has been gathered from plants in which the growing media had been amended with a particular trace element and/or a source of trace elements, such as sewage sludge, industrial wastes, etc. By comparison, little information is available on what would be considered the "background level" for the trace elements that exist in plants under normal growing conditions on soils that are neither deficient nor excessively high in their trace element

TABLE 1.40   The reagent ingredients for the A–Z Solutions, A and B

A		B	
Reagent	*mg/L*	*Reagent*	*mg/L*
$Al_2(SO_4)_3$	0.005	As_2O_3	0.0055
KI	0.027	$BaCl_2$	0.027
KBr	0.027	$CdCl_2$	0.0055
TiO_2	0.055	$Bi(NO_3)_3$	0.005
$SnCl_2 \cdot 2H_2O$	0.027	Rb_2SO_4	0.0055
LiCl	0.027	K_2CrO_4	0.027
H_3BO_3	0.61	KF	0.0055
$ZnSO_4 \cdot 7H_2O$	0.055	$PbCl_2$	0.0055
$CuSO_4 \cdot 5H_2O$	0.055	$HgCl_2$	0.055
$NiSO_4 \cdot 6H_2O$	0.055	H_2SeO_4	0.0055
$Co(NO_3)_2 \cdot 6H_2O$	0.055	$SrSO_4$	0.027
		VCl_3	0.0055

TABLE 1.41   Average concentrations of mineral nutrients in plant dry matter that are sufficient for adequate growth

Element	$\mu mol/g$ (dry weight)	mg/kg (ppm)	%	Relative Number of Atoms
Molybdenum (Mo)	0.001	0.1	—	1
Copper (Cu)	0.10	6	—	100
Zinc (Zn)	0.30	20	—	300
Manganese (Mn)	1.0	50	—	1,000
Iron (Fe)	2.0	100	—	2,000
Boron (B)	2.0	20	—	2,000
Chlorine (Cl)	3.0	—	0.1	3,000
Sulfur (S)	30	—	0.2	30,000
Phosphorus (P)	60	—	0.2	60,000
Magnesium (Mg)	80	—	0.2	80,000
Calcium (Ca)	125	—	0.5	125,000
Potassium (K)	250	—	1.0	250,000
Nitrogen (N)	1,000	—	1.5	1,000,000

Source: Epstein, 1965.

TABLE 1.42 Micronutrients essential for growth and their average content in material from cultivated higher plants and approximate concentration in the environment

Micronutrient	Mass Concentration (g/dry matter)	Concentration mmol (kg/dry matter)	No. of Atoms Relative to Molybdenum	Concentration in Environment (mol/m³)	Examples of Function in Cell
Chlorine (Cl)	0.1	3	3×10^3	0.001	Chloroplast photosystem II, metabolism, growth
Boron (B)	0.02	2	2×10^3	0.001	
Iron (Fe)	0.01	2	2×10^3	0.001	Energy transfer proteins, co-enzyme factor prosthetic groups
Manganese (Mn)	0.05	1	1×10^3	0.001	Co-factor in water splitting enzyme, aminopeptidase, etc.
Zinc (Zn)	0.02	0.3	3×10^3	7×10^{-4}	Enzyme co-factor, carbonic anhydrase, alkaline phosphatase, enzyme regulation
Copper (Cu)	0.06	0.1	1×10^3	3×10^{-4}	Constituent of plastocyanin, ascorbic acid oxidase, etc.
Molybdenum (Mo)	0.0001	0.001	1	5×10^{-4}	Constituent of nitrate reductase, nitrogenase

Source: Porter and Lawlor, 1991.

content (Davis et al., 1978; Markert, 1992). However, several investigators have suggested, based on their own determinations and/or literature values, what typical trace element contents would be in plants, mainly the leaves. For example, Melsted (1973) published his list of ranges and maximum values for a number of the trace elements for plant leaves in general and corn leaves in particular (Table 1.43).

TABLE 1.43   Normal range and suggested maximum trace element concentrations for plant leaves and suggested levels for corn leaves

| | Concentration of Metals in Plants (mg/kg, dry wt) | | | |
| | Plant Leaves | | Corn Leaves | |
Element	Range	Maximum	Range	Maximum
Arsenic (As)	0.01–1.0	2	Same	Same
Barium (Ba)	10–100	200	Same	Same
Boron (B)	7–75	150	Same	100
Cadmium (Cd)	0.05–0.20	3	0.05–0.20	1–3[a]
Chromium (Cr)	0.1–0.5	2	0.05–1.0	5
Cobalt (Co)	0.01–0.30	5	Same	Same
Copper (Cu)	3–40	150	5–25	30
Fluorine (F)	1–5	10	1–10	50
Iodine (I)	0.1–0.5	1	Same	Same
Iron (Fe)	20–300	750	50–200	300
Lead (Pb)	0.1–5.0	10	Same	Same
Lithium (Li)	0.2–1.0	5	Same	Same
Manganese (Mn)	15–150	300	Same	Same
Mercury (Hg)	0.001–0.01	0.04	Unknown[b]	
Molybdenum (Mo)	0.2–1.0	3	0.2–1.0	5[c]
Nickel (Ni)	0.1–1.0	3	0.1–10	20
Selenium (Se)	0.05–2.0	3	0.1–5	10
Vanadium (V)	0.1–1.0	2	Same	Same
Zinc (Zn)	15–150	300	20–100	300

[a] The level in silage corn must be maintained at a lower level of Cd than for corn grain.

[b] Without more experience, the values of Melsted could be used, but organic complexes of Hg could provide a threat to animal health at very low levels of total Hg.

[c] The level of Mo which is toxic to cattle depends upon the concentration of Cu in rations. The recommended maximum could be toxic in very low Cu rations.

Source: Melsted, 1973.

TABLE 1.44 Trace element content of "reference plant"

Trace Element	mg/kg	Trace Element	mg/kg
Aluminum (Al)	80	Iodine (I)	3.0
Antimony (Sb)	0.1	Iron (Fe)	150
Arsenic (As)	0.1	Lead (Pb)	1.0
Barium (Ba)	40	Manganese (Mn)	200
Beryllium (Be)	0.001	Mercury (Hg)	0.1
Bismuth (Bi)	0.01	Molybdenum (Mo)	0.5
Boron (B)	40	Nickel (Ni)	1.5
Bromine (Br)	4.0	Selenium (Se)	0.02
Cadmium (Cd)	0.05	Silver (Ag)	0.2
Cerium (Ce)	0.5	Strontium (Sr)	50
Cesium (Cs)	0.2	Thallium (Tl)	0.05
Chromium (Cr)	1.5	Tin (Sn)	0.2
Cobalt (Co)	0.2	Titanium (Ti)	5.0
Copper (Cu)	10	Tungsten (W)	0.2
Fluorine (F)	2.0	Uranium (U)	0.01
Gallium (Ga)	0.1	Vanadium (V)	0.5
Gold (Au)	0.001	Zinc (Zn)	50

Note: No data from typical accumulator and/or rejector plants.

Source: Markert, 1994a.

More recently, Markert (1994a) published his listing of the trace element content of a "reference plant" and what might be considered as "normal" for the trace element content of plants in general (Table 1.44).

Soil and plant contents for a number of the trace elements are given in Table 1.45, indicating the range and common concentration for soils and the normal and toxic levels for plants.

The tabular data given in Tables 1.43 to 1.45 are trace element concentrations that could be considered as that expected to be found in normally growing plants, influenced by neither trace element concentration extremes nor deficiency or toxicity in the rooting environment. However, when a trace element level in the rooting environment is either deficient or excessive, the concentration level in the plant may, in turn, significantly affect its growth and development. A generalized listing of concentration ranges, from deficiency to toxicity, for the trace elements in mature leaves is given in Table 1.46.

TABLE 1.45   Total concentration of various elements typically found in soil and plants

	μg/g			
	Soils		Plant	
Element	Common Range	Level	Normal	Toxic[a]
Arsenic (As)	0.1–40	6	0.1–5	—[b]
Boron (B)	2.0–100	10	30–75	>75
Cadmium (Cd)	0.01–7	0.06	0.2–0.8	>2
Chromium (Cr)	5.0–3000	100	0.2–1	—
Cobalt (Co)	1.0–400	8	0.05–0.5	—
Copper (Cu)	2.0–100	20	4–15	>20
Lead (Pb)	2.0–200	10	0.1–10	—
Manganese (Mn)	100–4000	850	15–100	—
Molybdenum (Mo)	0.2–5	2	1.0–100	—
Nickel (Ni)	10–1000	40	1	50
Selenium (Se)	0.1–2	0.5	0.02–2	50–100
Vanadium (V)	20–500	100	0.1–10	>10
Zinc (Zn)	10–300	50	15–200	>200

[a] Toxicities listed do not apply to certain accumulator plants.
[b] — = no data.

Deficiency symptoms for the essential micronutrients have been identified and described in considerable detail for many crops (see Chapter 4). Although much of the focus regarding the micronutrients relates to deficiency, there is growing interest in identifying symptoms related to toxic (frequently referred to as phytotoxicity—the ability of a trace element to kill or seriously harm a plant) concentrations of the trace elements in the plant (Beckett and Davis, 1977). The most common and non-specific symptoms of phytotoxicity are

- Chlorotic or brown points on leaves and leaf margins
- Brown, stunted coralloid roots

However, a more detailed description of toxicity symptoms and the most sensitive crops are given in Table 1.47.

Basic reactions related to the toxic effects of excessive amounts of the trace elements on plants have been described as follows:

TABLE 1.46 Approximate concentrations of trace elements in mature leaf tissue generalized for various species

Trace Element	*mg/kg dry weight*		
	Deficient or Normal	*Sufficient or Toxic*	*Excessive*
Antimony (Sb)	–	7–50	150
Arsenic (As)	–	1–1.7	5–20
Barium (Ba)	–	–	500
Beryllium (Be)	–	<1–7	10–50
Boron (B)	5–30	10–200	50–200
Cadmium (Cd)	–	0.05–0.2	5–30
Chromium (Cr)	–	0.1–0.5	5–30
Cobalt (Co)	–	0.02–1	15–50
Copper (Cu)	2–5	5–30	20–100
Fluorine (F)	–	5–30	50–500
Lead (Pb)	–	5–10	30–300
Lithium (Li)	–	3	5–50
Manganese (Mn)	15–25	20–300	300–500
Mercury (Hg)	–	–	1–3
Molybdenum (Mo)	0.1–0.3	0.2–1	10–50
Nickel (Ni)	–	0.1–5	10–100
Selenium (Se)	–	0.001–2	5–30
Silver (Ag)	–	0.5	5–10
Thallium (Tl)	–	–	20
Tin (Sn)	–	–	60
Titanium (Ti)	0.2–0.5	0.5–2.0	50–200
Vanadium (V)	–	0.2–1.5	5–10
Zinc (Zn)	10–20	27–150	100–400
Zirconium (Zr)	0.2–0.5	0.5–2.0	15

Source: Kabata-Pendias and Pendias, 1994.

Trace Elements	*Reaction*
Bromine, cadmium, copper, fluorine, gold, iron, mercury, lead, silver	Changes in permeability of the cell membrane
Lead, mercury, silver	Reactions of thiol groups with cations
Antimony, arsenic, fluorine, selenium, tellurium, tungsten	Competition for sites with essential metabolites

TABLE 1.47   General effects of trace element toxicity on common cultivars

Trace Element	Symptoms	Sensitive Crops
Aluminum (Al)	Overall stunting, dark green leaves, purpling of stems, death of leaf tips, and coralloid and damaged root system.	Cereals
Arsenic (As)	Red-brown necrotic spots on old leaves, yellowing and browning of roots, depressed tillering wilting of new leaves.	Legumes, onion, spinach, cucumbers, bromegrass, apricots, peaches
Beryllium (Be)	Inhibition of seed germination and reduced growth, degradation of protein enzymes.	—
Boron (B)	Margin or leaf tip chlorosis, browning of leaf points, decaying growing points, and wilting and dying-off of older leaves. In severely affected pine trees, necrosis occurs on needles near the ends of shoots and in upper half of the tree.	Cereals, potatoes, tomatoes, cucumbers, sunflowers, mustard, apple, apricots, citrus, walnut
Cadmium (Cd)	Brown margin of leaves, chlorosis, reddish veins and petioles, curled leaves, and brown stunted roots. Severe reduction in growth of roots, tops, and number of tillers (in rice). Reduced conductivity of stem, caused by deterioration of xylem tissues.	Legumes (bean, soybean), spinach, radish, carrots
Chromium (Cr)	Chlorosis of new leaves, necrotic spots and purpling, tissues injured root growth.	—
Cobalt (Co)	Interveinal chlorosis in new leaves followed by induced Fe chlorosis and white leaf margins and tips, and damaged root tips.	—
Copper (Cu)	Dark green leaves followed by induced Fe chlorosis; thick, short, or barbed-wire roots; depressed tillering.	Cereals and legumes, spinach, citrus seedlings, gladiolus
Fluorine (F)	Margin and leaf tip necrosis, and chlorotic and red-brown points of leaves.	Gladiolus, grapes, fruit trees, pine trees
Iron (Fe)	Dark green foliage, stunted growth of tips and roots, dark brown to purple leaves of some plants (e.g., "bronzing" disease of rice).	Rice and tobacco

TABLE 1.47 **General effects of trace element toxicity on common cultivars (continued)**

Trace Element	Symptoms	Sensitive Crops
Lead (Pb)	Dark green leaves, wilting of older leaves, stunted foliage, and brown short roots.	—
Manganese (Mn)	Chlorosis and necrotic lesions on old leaves, blackish or red necrotic spots, accumulation of MnO particles in epidermal cells, drying tips of leaves, and stunted roots and plant growth.	Cereals, legumes, potatoes, cabbage
Mercury (Hg)	Severe stunting of seedlings and roots, leaf chlorosis, and browning of leaf points.	Sugarbeet, maize, roses
Molybdenum (Mo)	Yellowing or browning of leaves, depressed root growth, depressed tillering.	Cereals
Nickel (Ni)	Interveinal chlorosis (caused by Fe-induced deficiency in new leaves), gray-green leaves, and brown and stunted roots and plant growth.	Cereals
Rubidium (Rb)	Dark green leaves, stunted foliage, and increasing amount of shoots.	—
Selenium (Se)	Interveinal chlorosis or black spots at Se content of about 4 ppm, and complete bleaching or yellowing of younger leaves at higher Se content, pinkish spots on roots.	—
Thallium (Tl)	Impairment of chlorophyll synthesis, mild chlorosis and slight cupping of leaves, reduced germination of seeds and growth of plants.	Tobacco and cereals
Titanium (Ti)	Chlorosis and necrosis of leaves, stunted growth.	Beans
Zinc (Zn)	Chlorotic and necrotic leaf tips, interveinal chlorosis in new leaves, retarded growth of entire plant, and injured roots resemble barbed wire.	Cereals and spinach

Source: Kabata-Pendias and Pendias, 1994.

Trace Elements	*Reaction*
Aluminum, beryllium, scandium, yttrium, zirconium	Affinity for reaction with phosphate groups and active groups of ADP or ATP
Cesium, lithium, rubidium, selenium, strontium	Replacement of essential ions (mainly major cations)
Arsenate, borate, bromate, fluorate, selenate, tellurate, and tungstate	Occupation of sites for essential groups such as phosphate and nitrate

Plants themselves have the ability to adjust to trace element excesses. Some of the possible mechanisms involved in metal tolerance are

- Selective uptake of ions
- Decreased permeability of membranes or other differences in the structure and function of membranes
- Immobilization of ions in roots, foliage, and seeds
- Removal of ions from metabolism by deposition (storage) in fixed and/or insoluble forms in various organs and organelles
- Alterations in metabolic patterns—increased enzyme system that is inhibited, or increased antagonistic metabolite, or reduced metabolic pathway by passing an inhibited site
- Adaptation to toxic metal replacement of a physiological metal in an enzyme
- Release of ions from plants by leaching from foliage, guttation, leaf shedding, and excretion from roots

In addition, trace element toxicity correlation factors can affect the sensitivity of plants to a particular trace element:

- Electronegativity of divalent metals
- Solubility products of sulfides
- Stability of chelates
- Bioavailability

Toxicity and/or tolerance of a trace element by plants can occur due to an elemental interaction, mainly with a major element, as well as antagonistic or synergistic effects, some of which are given in Table 1.48. Similar interelement correlations based on reference materials and research samples have been discussed by Markert (1993).

Wallace (1971) has suggested that one of the major roles of calcium in the plant is to counter the toxicity of so-called heavy metals (some being essential micronutrients) that may exist at high concentrations in the plant at levels that can be toxic. The calcium concentration required to counter a

TABLE 1.48 Interactions between major elements and trace elements in plants

Major Element	Antagonistic Elements	Synergistic Elements
Calcium	Aluminum, barium, beryllium, boron, cadmium, cesium, chromium, cobalt, copper, fluorine, iron, lead, lithium, manganese, nickel, strontium, zinc	Copper, manganese, zinc
Magnesium	Aluminum, barium, beryllium, chromium, cobalt, copper, fluorine, iron, manganese, nickel, zinc	Aluminum, zinc
Phosphorus	Aluminum, arsenic, boron, beryllium, cadmium, chromium, copper, fluorine, iron, lead, manganese, mercury, molybdenum, nickel, rubidium, scandium, silicon, strontium, zinc	Aluminum, boron, copper, fluorine, iron, manganese, molybdenum, zinc
Potassium	Aluminum, boron, cadmium, chromium, fluorine, manganese, mercury, molybdenum, rubidium	
Sulfur	Arsenic, barium, iron, lead, molybdenum, selenium	Fluorine, iron
Nitrogen	Boron, copper, fluorine	Boron, copper, iron, molybdenum
Chlorine	Bromine, iodine	

trace element toxicity has been suggested to be 2% or greater. The physiology of metal toxicity has been discussed by Foy et al. (1978) and critical tissue concentrations by MacNicol and Beckett (1985).

The soil, the common rooting medium, is the primary source for the trace elements found in plants. In general, as the trace element content of the soil increases, that amount available to the plant also increases, although there are other soil factors (such as pH, level of organic matter, texture, etc.) that will determine what portion of the trace element content of a soil will be available for root absorption (De Temmerman et al., 1984). The range, common level, and amount of trace elements in the soil that is tolerable to plants are given in Table 1.49. The trace element content of agricultural soils has recently been reviewed by Logan and Traina (1993).

TABLE 1.49   Amount of various trace elements tolerable to plants

	mg/kg		
Element	Range	Common Level	Amount Tolerable (proposed)
Arsenic (As)	1.0–50	2.0–20	50
Beryllium (Be)[a]	0.1–10	1.0–5	10
Boron (B)	2.0–100	5.0–30	100
Cadmium (Cd)[a]	0.01–1.0	0.1–1.0	5
Chromium (Cr)	1.0–100	10–50	100
Cobalt (Co)	1.0–50	1.0–10	50
Copper (Cu)	2.0–100	5.0–20	100
Fluorine (F)	10–500	50–250	500
Lead (Pb)	0.1–10	0.1–5	100
Mercury (Hg)[a]	0.01–1.0	0.1–1.0	5
Molybdenum (Mo)	0.2–10	1.0–5	10
Nickel (Ni)	1.0–100	10–50	100
Selenium (Se)	0.1–10	1.0–5	10
Zinc (Zn)	10–300	10–50	300

[a] The effect these elements have on plants has not been determined.

A number of trace elements have been of major concern regarding crop production effects as well as their potential introduction into the food chain (Berrow and Burridge, 1979; Bingham et al., 1986). A recent study conducted by Sillanpää and Jansson (1992) focused on the levels of cadmium, lead, cobalt, and selenium found in soils and two major grain crops, maize and wheat. Soil and plant tissue samples were collected from 30 different countries in various parts of the world. The results provide useful information on the effect of plant type on uptake, range in concentration found, and the relationship between soil test levels and that found in the plant.

The authors also gathered data on the content of these 4 trace elements in 17 different crops in side-by-side growth comparisons. The results are given in Table 1.50.

The heavy metal content of some plants may be significantly influenced by their ability to be adsorbed and accumulate in a particular plant part, as shown in Table 1.51. Such characteristics are useful for selecting those crop plants best suited for soils that have a heavy metal content, thereby avoiding

TABLE 1.50 **Average contents of four trace elements in the dry matter of different crops grown side by side**

Crop and Part Sampled	Cadmium (ppm)	Lead (ppm)	Cobalt (ppm)	Selenium (ppb)
Spring wheat				
Grain	0.089 ± 0.064	0.06 ± 0.02	0.033 ± 0.042	4.7 ± 0.8
Straw	0.258 ± 0.174	0.67 ± 0.37	0.049 ± 0.051	6.1 ± 2.3
Winter wheat				
Grain	0.058 ± 0.017	0.07 ± 0.03	0.041 ± 0.040	—
Straw	0.253 ± 0.153	0.46 ± 0.22	0.056 ± 0.063	—
Oats				
Grain	0.026 ± 0.018	0.10 ± 0.04	0.027 ± 0.023	5.7 ± 1.3
Straw	—	—	0.56 ± 0.28	0.078 ± 0.080
Barley				
Grain	0.032 ± 0.024	0.10 ± 0.06	0.021 ± 0.016	4.8 ± 0.9
Straw	—	—	0.53 ± 0.22	0.054 ± 0.046
Rye				
Grain	—	—	0.20 ± 0.17	0.047 ± 0.068
Straw	0.130 ± 0.069	0.50 ± 0.30	0.064 ± 0.081	—
Timothy, silage				
First cut	0.034 ± 0.015	0.40 ± 0.28	0.068 ± 0.089	6.5 ± 2.6
Second cut	0.040 ± 0.019	0.64 ± 0.42	0.110 ± 0.120	8.5 ± 3.5
Timothy				
Dry hay	0.035 ± 0.023	0.42 ± 0.27	0.061 ± 0.068	6.2 ± 3.2
Fresh growth	0.058 ± 0.029	1.14 ± 0.86	0.084 ± 0.059	11.0 ± 4.9
Ryegrass, silage				
First cut	0.079 ± 0.032	0.70 ± 0.45	0.22 ± 0.37	9.3 ± 3.6
Second cut	0.103 ± 0.042	1.55 ± 1.86	0.29 ± 0.56	12.4 ± 3.4
Red clover				
Dry hay	0.083 ± 0.084	0.63 ± 0.25	0.53 ± 0.81	6.0 ± 3.1
Fresh growth	0.108 ± 0.095	1.13 ± 1.04	0.55 ± 1.02	9.3 ± 4.3
Pea				
Seed	0.007 ± 0.006	0.13 ± 0.05	0.11 ± 0.21	6.4 ± 3.3
Stalk	0.135 ± 0.140	1.19 ± 0.77	0.38 ± 0.58	—
Turnip rape				
Seed	0.080 ± 0.029	0.27 ± 0.06	0.12 ± 0.17	8.1 ± 2.8
Stalk	0.341 ± 0.141	0.59 ± 0.53	0.20 ± 0.40	7.9 ± 3.7
Turnip, root	0.167 ± 0.076	0.15 ± 0.09	0.20 ± 0.33	—
Swede				
Root	0.096 ± 0.041	0.21 ± 0.19	0.29 ± 0.48	7.4 ± 2.1
Tops	0.275 ± 0.177	1.28 ± 0.81	0.24 ± 0.35	20.1 ± 9.2

TABLE 1.50   Average contents of four trace elements in the dry matter of different crops grown side by side (continued)

Crop and Part Sampled	Cadmium (ppm)	Lead (ppm)	Cobalt (ppm)	Selenium (ppb)
Sugar beet				
Root	0.203 ± 0.083	0.14 ± 0.08	0.10 ± 0.08	—
Tops	0.687 ± 0.281	1.25 ± 0.98	0.24 ± 0.16	—
Red beet, root	0.337 ± 0.124	0.09 ± 0.06	0.27 ± 0.27	5.0 ± 0.8
Carrot, root	0.431 ± 0.145	0.25 ± 0.38	0.12 ± 0.19	7.8 ± 2.1
Potato, tuber	—	—	0.05 ± 0.03	0.09 ± 0.11
Onion, bulb	0.217 ± 0.090	0.21 ± 0.17	0.10 ± 0.10	6.9 ± 1.4

Note: Mean ± standard deviation.

Source: Ylärnta, 1990; Sillanpää and Jansson, 1992.

the potential of introducing a heavy metal into the food chain (Chaney, 1980, 1983; Chaney et al., 1987).

Not all heavy metals have the same soil–plant characteristics in terms of their ability to be taken up by the plant. Alloway (1995) determined the soil–plant transfer coefficients which are given in Table 1.52.

Based on the accumulation by plants for various trace elements, Kabata-Pendias and Pendias (1994) determined crop removal levels for eight trace elements by reference and accumulator plants, giving the percentage of

TABLE 1.51   Relative metal accumulation (cadmium and lead in the edible portions; copper, nickel, and zinc in the leaves)

Element	High Accumulators	Low Accumulators
Cadmium (Cd)	Lettuce, spinach, celery, cabbage	Potato, maize, french bean, peas
Copper (Cu)	Sugar beet, certain barley cultivars	Leek, cabbage, onion
Lead (Pb)	Kale, ryegrass, celery	Some barley cultivars, potato, maize
Nickel (Ni)	Sugar beet, ryegrass, marigold, onion	Maize, leek, barley cultivars, turnip
Zinc (Zn)	Sugar beet, marigold, spinach, beetroot	Potato, leek, tomato, onion

TABLE 1.52 Soil–plant transfer coefficients for 13 heavy metals

Element	Soil–Plant Transfer Coefficient
Arsenic (As)	0.01–0.1
Beryllium (Be)	0.01–0.1
Cadmium (Cd)	1–10
Chromium (Cr)	0.01–0.1
Cobalt (Co)	0.01–0.1
Copper (Cu)	0.1–10
Lead (Pb)	0.01–0.1
Mercury (Hg)	0.01–0.1
Nickel (Ni)	0.1–1.0
Selenium (Se)	0.1–10
Thallium (Tl)	1–10
Tin (Sn)	0.01–0.1
Zinc (Zn)	1–10

Source: Alloway, 1995.

elements taken up by the plant based on the content in the soil. The percentage accumulated varied from 0.06 to 0.5% for the reference crops and 0.3 to 10% for the accumulator plants, suggesting that very little of the soil content is accumulated by most crop plants (Table 1.53).

TABLE 1.53 Trace metal removal by crops from soil

Trace Element	Content of Soil	Output with Plant			
		Reference		Accumulator	
	kg/ha	g/ha	%[a]	g/ha	%[a]
Cadmium (Cd)	1.5	1	0.06	100	10.0
Chromium (Cr)	150	50	0.03	500	0.3
Copper (Cu)	45	100	0.2	500	1.0
Lead (Pb)	75	100	0.1	500	0.6
Manganese (Mn)	810	1000	0.1	5000	0.6
Molybdenum (Mo)	6	30	0.5	250	4.0
Nickel (Ni)	39	50	0.1	100	0.3
Zinc (Zn)	135	400	0.3	1500	1.0

[a] As percent of the total content of soil.

Source: Kabata-Pendias and Pendias, 1994.

ANIMAL AND HUMAN PHYSIOLOGY

The trace element content of animal and human tissues, in general, is not as well known as that for plants, nor have the trace elements been as intensively studied by animal and human physiologists as has been the case for plants (Skoryna et al., 1989; van Campen, 1991). It should also be noted that the physical body and body fluids in animals and humans are far more complex and varied than that of plants. The history associated with the determination of the elements found in plants and animal bodies has many parallels in terms of scientific discovery. McDowell (1992) prepared a chronological sequence of discovery related to the mineral elements affecting animal and human health, as shown in Table 1.54. This compares with the information given in Table 1.39, which gives the date of discovery and discoverer for those micronutrients essential for plants.

TABLE 1.54   History of nutritional importance of mineral elements

Time	Event
29 B.C.	The fall of Thebes was hastened by heavy livestock mortalities (caused by unidentified agents) while grazing luxuriant pastures.
40–120 A.D.	Salt fed to domestic animals during the time of Plutarch.
23–79 A.D.	Virgil and Pliny recommended salts for milk production.
1295	Clinical signs of selenium toxicity were apparently described by Marco Polo as affecting grazing livestock in China.
before 1680	Sydenham treated anemia with iron filings.
1747	Menghini found iron in blood.
1748	Gahn reported phosphorus present in bones.
1770	Scheele reported that bones contain calcium phosphate.
1784	Scheele reported sulfur in proteins.
1791	Fordyce showed that canaries need "calcareous earth" supplements to grain diets.
1811–1825	Work by Courtois, Coindet, and Boussingault led to the discovery of iodine, the effectiveness of iodine in burnt sponges, and specifically that iodine was the only cure for goiter.
1823	Proust reported chlorine in the hydrochloric acid in gastric juice.
1842	Chossat found that pigeons required calcium for bone growth.

TABLE 1.54 **History of nutritional importance of mineral elements (continued)**

Time	Event
1847	Liebig reported potassium in animal tissues.
1850–1854	Chatin published studies relating environmental iodine deficiency to incidence of endemic goiter in man and animals.
1869	Raulin discovered the essentiality of zinc for the microorganism *Aspergillus niger.*
1873	Von Bunge put forth the hypothesis of antagonism between sodium and potassium and between sodium and chlorine.
1880	Forster demonstrated that animals require minerals and that feeding dogs only meat resulted in deficiencies.
1893–1899	Von Bunge and Abderhalden showed that young animals receiving milk require supplemental iron.
1905	Babcock studied salt requirements of cattle, noting its particular importance for lactating cows.
1919	Kendall isolated and named thyroxin from thyroid gland; the hormone was found to contain 65% iodine.
1920	Bertrand in France and McHargue in the United States initiated the use of purified diets to study the need for and function of various minerals.
1922	Bertrand and Berzon showed that zinc is necessary for rat growth and hair development.
1924	Theiller and co-workers studied phosphorus deficiency for grazing cattle and found that supplementation corrects bone chewing, prevents death loss from botulism, and increases growth and reproductive rates.
1926	Leroy showed that magnesium increases growth in mice.
1928	Hart and co-workers showed that copper, in addition to iron, is needed for hemoglobin formation.
1928–1933	Warburg established that respiratory enzymes in animals contain an iron porphyrin group.
1931	Neal, Becker, and Shealy established copper as an essential element for ruminants.
1931	Kemerer and McCollum showed that manganese is essential for rats and mice and that a deficiency causes tetany.
1933	Sjollema related a licking disease in cattle to copper deficiency.

TABLE 1.54   History of nutritional importance of mineral elements (continued)

Time	Event
1935	Franke and Potter identified selenium as the factor in forage responsible for alkali disease in farm animals.
1935	Duncan and Huffman observed tetany in calves due to low magnesium content of milk.
1935	Underwood and Filmer and, independently, Marston and Lines found that enzootic marasmus in sheep is caused by a cobalt deficiency.
1936–1937	Wilgus, Norris, and Houser reported that manganese deficiency results in a perosis in chicks.
1937	Becker and co-workers established that the "salt sick" condition of cattle in Florida is caused by a combination of pasture deficiencies of cobalt, copper, and iron.
1937	Bennets and Chapman demonstrated that enzootic ataxia of newborn lambs resulted from ewes receiving insufficient copper during pregnancy.
1938	Ferguson, Lewis, and Watson showed that molybdenum toxicity results in severe diarrhea in grazing cattle.
1938–1942	Hevesy and others began to use radioisotopes to study mineral metabolism.
1940	Leilin and Mann reported zinc as a component of the enzyme carbonic anhydrase.
1946	Moulton established that small concentrations of fluorine in drinking water prevent dental caries.
1948	Rickes and co-workers and, independently, Smith showed that cobalt is an integral part of vitamin B_{12}.
1950–1954	Dick noted metabolic interrelationships among copper, molybdenum, and inorganic sulfates in ruminants.
1953	Richert and Westerfield isolated molybdenum from the metalloenzyme xanthine oxidase.
1954	Needy and Harbaugh found that high fluorine concentrations in drinking water result in mottling of tooth enamel.
1955	Tucker and Salmon discovered that parakeratosis, a severe skin disease, is caused by a zinc deficiency in swine.
1957	Schwartz and Foltz identified selenium as a factor that prevents liver necrosis in rats.

TABLE 1.54 History of nutritional importance of mineral elements (continued)

Time	Event
1958–1959	Scott prevented exudate diathesis in poultry with selenium, while Muth, Oldfield, Remmert, McLean, Thompson, Claxton, and others prevented white muscle disease in ruminants with this element.
1959	Schwartz and Mertz showed that chromium was essential for glucose metabolism.
1970–1984	The most recently discovered elements (new trace elements) were established using highly purified diets and metal-free isolator systems. These elements included arsenic, boron, lead, lithium, nickel, silicon, tin, and vanadium.

Source: McDowell, 1992.

The rigid requirements of essentiality established by Arnon and Stout (1939) for plants could be accepted as valid for animals and humans, but some years ago the criteria for essentiality were enlarged to include the following:

- The element is present in tissues of different animals at comparable concentrations.
- Its withdrawal produces similar physiological or structural abnormalities regardless of species.
- Its presence reverses or prevents these abnormalities.
- These abnormalities are accompanied by specific biochemical changes that can be remedied or prevented when the deficiency is checked.

These criteria were published by Horovitz (1988) in a paper entitled "Is the Major Part of the Periodic System Really Essential for Life?"

The relative concentrations of the elements found in an adult human body are shown in Figure 1.4.

For animals and humans to develop normally and sustain health, the following trace elements have been identified as being "essential": copper, iron, manganese, molybdenum, zinc, selenium, chromium, iodine, fluorine, and cobalt; the trace elements lithium, nickel, arsenic, silicon, tin, and vanadium have been classed as "beneficial" (see Table 1.37). Both lead and cadmium have been shown to affect animal growth and development (hav-

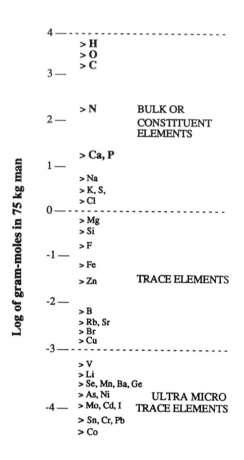

FIGURE 1.4 Elemental composition of the human adult expressed on a logarithmic scale. (Source: Frieden, 1981.)

ing some degree of beneficial character) when at very low concentrations, although these two elements are usually considered highly toxic at relatively moderate concentrations in body tissues; however, such suggestions of beneficiality are not widely accepted. In addition, published reports with convincing data suggest that rubidium, bromine, titanium, and some of the rare earth metals may have beneficial and/or stimulative effects on animals, but again, for full acceptance of these reported observations, additional studies will be required.

The uptake of trace elements by animal and human organisms is more complicated than in the case of plants (McDowell, 1989). Chelating agents in various animal and human cells can assist (e.g., vitamin C) in uptake, as can a wide spectrum of organic compounds (e.g., phytic acid) which will hinder the bioavailability of various trace elements. Therefore, a trace element content determination of an ingested food or feed will not estimate intake correctly, whereas the assay of widely accepted "indicator organs," such as covering hair or fingernails, and the direct analysis of blood serum or urine are much better evaluators of the trace element status of the animal or human body. The trace element content of these types of body tissues and fluids is given in Tables 1.55 to 1.57.

Relatively little data is available on the trace element content of animals, whether specific tissues or fluids, although whole animal trace element contents have been published, as shown in Table 1.58.

TABLE 1.55 Urine and scalp hair element concentration parameters

	Urine (n = 300)			Scalp Hair (adult females, n = 960)		
	mg/L			*mg/kg*		
Trace Element	*Mean*	*1 SD Low*	*1 SD High*	*Mean*	*1 SD Low*	*1 SD High*
Aluminum (Al)	—	0.040	—	3.147	0.813	12.17
Arsenic (As)	0.144	0.077	0.269	1.574	0.654	3.791
Barium (Ba)	—	0.005	—	1.09	0.338	3.50
Beryllium (Be)	—	0.005	—	0.036	0.021	0.060
Boron (B)	2.48	1.54	4.02	1.44	0.906	2.64
Cadmium (Cd)	—	0.020	—	0.354	0.176	0.709
Chromium (Cr)	—	0.030	—	0.491	0.345	0.864
Cobalt (Co)	—	0.020	—	0.195	0.121	0.315
Copper (Cu)	—	0.003	—	26.0	13.4	50.2
Gold (Au)	—	0.020	—	0.174	0.064	0.476
Iron (Fe)	0.081	0.029	0.227	10.1	6.02	17.1
Lead (Pb)	—	0.040	—	4.765	2.334	9.729
Manganese (Mn)	—	0.005	—	0.715	0.309	1.65
Mercury (Hg)	—	0.080	—	0.974	0.561	1.688
Selenium (Se)	0.010	0.004	0.025	0.384	0.166	0.887
Silver (Ag)		0.020	—	0.244	0.105	0.564
Strontium (Sr)	0.104	0.046	0.232	2.83	0.761	10.5
Vanadium (V)	—	0.030	—	0.156	0.090	0.269
Zinc (Zn)	0.333	0.139	0.793	154	128	184
Zirconium (Zr)	—	0.040	—	0.290	0.126	0.663

Source: Bederka et al., 1985.

TABLE 1.56 The trace element content of various human tissues

Trace Element	Muscle (mg/kg)	Bone (mg/kg)	Blood (mg/dm³)	Daily Dietary Intake (mg)	Toxic Intake (g)	Lethal Intake (mg)
Aluminum (Al)	0.7–28	4–27	0.39	2.45	5.0	unknown
Antimony (Sb)	0.042–0.191	0.01–0.6	0.0033	0.002–1.3	0.1	unknown
Arsenic (As)	0.009–0.65	0.08–1.6	0.0017–0.09	0.04–1.4	50–340	3,700
Barium (Ba)	0.09	3–70	0.068	0.60–1.7	0.2	unknown
Beryllium (Be)	0.00075	0.003	$<1 \times 10^{-5}$	0.01	unknown	unknown
Bismuth (Bi)	0.032	<0.2	~0.016	0.005–0.02	unknown	unknown
Boron (B)	0.33–1	1.1–3.3	0.13	1–3	4.0	unknown
Cadmium (Cd)	0.14–3.2	1.8	0.0052	0.007–3	0.003–0.330	unknown
Cesium (Cs)	0.07–1.6	0.013–0.052	0.0038	0.004–0.03	non-toxic	unknown
Chromium (Cr)	0.024–0.84	0.1–33	0.006–0.11	0.01–1.2	0.2	>3,000
Cobalt (Co)	0.028–0.65	0.01–0.04	0.0002–0.04	0.005–1.8	0.5	unknown
Copper (Cu)	10	1–26	1.01	0.50–6	0.25	unknown
Fluorine (F)	0.05	200–12,000	0.5	0.3–0.5	0.02 (as F⁻)	2,000 (as F⁻)
Gallium (Ga)	0.0014	unknown	<0.08	unknown	low toxicity	unknown
Germanium (Ge)	0.14	unknown	~0.44	0.4–1.5	non-toxic	unknown
Gold (Au)	unknown	0.016	0.1–4.2	very low	non-toxic	unknown
Indium (In)	0.015	unknown	unknown	low	0.030	>200
Iodine (I)	0.05–0.5	0.27	0.057	0.1–0.2	0.002	35–350
Iridium (Ir)	2×10^{-5}?	unknown	very low	very low	low toxicity	unknown
Iron (Fe)	180	3–380	447	6–40	0.20	7–35
Lanthanum (La)	0.004	<0.08	unknown	very low	unknown	720 (rats)
Lead (Pb)	0.23–3.3	3.6–30	0.21	0.06–0.5	0.001	10,000
Lithium (Li)	0.023	unknown	0.004	0.1–2.0	0.092–0.2	unknown
Manganese (Mn)	0.2–2.3	0.2–100	0.0016–0.075	0.4–10	0.01–0.02 (rats)	unknown

Element						
Mercury (Hg)	0.02–0.7	0.45	0.0078	0.004–0.02	0.0004	150–300
Molybdenum (Mo)	0.018	<0.7	~0.001	0.05–0.35	0.005	50 (rats)
Nickel (Ni)	1–2	<0.7	0.001–0.05	0.3–0.5	0.050 (rats)	unknown
Niobium (Nb)	0.14	<0.07	0.005?	0.02–0.6	moderately toxic	unknown
Platinum (Pt)			unknown but low			
Rubidium (Rb)	20–70	0.1–5	2.49	1.5–6	non-toxic	unknown
Scandium (Sc)	unknown	~0.001	~0.008	~0.00005	low toxicity	unknown
Selenium (Se)	0.42–1.9	1–9	0.171	0.06–0.2	0.005	unknown
Silicon (Si)	100–200	17	3.9	18–1200	non-toxic	unknown
Silver (Ag)	0.009–0.28	0.01–0.44	<0.003	0.0014–0.08	0.060	1.3–6.2
Strontium (Sr)	0.12–0.35	36–140	0.031	0.8–5.0	non-toxic	unknown
Tantalum (Ta)	unknown	~0.03	unknown	0.001	unknown	300 (rats)
Tellurium (Te)	0.017?	unknown	0.0055?	<0.1	0.00025	2
Thallium (Tl)	0.07	0.002	0.00048	0.0015	unknown	600
Thorium (Th)	unknown	0.002–0.012	0.00016	0.00003–0.003	low toxicity	unknown
Tin (Sn)	0.33–2.4	1.4	~0.38	0.2–3.5	2,000	unknown
Titanium (Ti)	0.9–2.2	unknown	0.0054	0.8	low toxicity	unknown
Tungsten (W)	unknown	0.00025	0.001	0.001–0.015	unknown	>30 (rats)
Uranium (U)	9×10^{-4}	$(0.016\text{–}70) \times 10^{-3}$	5×10^{-4}	0.001–0.002	unknown	30 (rats)
Vanadium (V)	0.02	0.0035	<0.0002	0.04	0.00025	2–4
Ytterbium (Yb)			unknown but low			
Yttrium (Y)	0.02	0.07	0.0047	0.016	low toxicity	unknown
Zinc (Zn)	240	75–170	7.0	5–40	0.15–0.6	6
Zirconium (Zr)	0.08	<0.1	0.011	~0.05	non-toxic	unknown

Source: Emsley, 1991.

TABLE 1.57 Trace element content of human body fluids and tissues

Trace Element	Kidney (μg/g)	Milk (mg/L)	Tooth (μg/g)	Urine (μg/L)
Aluminum (Al)	0.4	0.33	8–325	3.5–31
Antimony (Sb)	1.0–42	<0.2	0.02–0.34	<1.0
Arsenic (As)	0.45–82	1.6–6	<0.02	<10
Barium (Ba)	0.01 ± 0.001	20	129 ± 54.7	4.8 ± 1.4
Beryllium (Be)	—	—	—	0.4–0.9
Bismuth (Bi)	0.4 ± 0.04	—	—	<1.0–2.4
Boron (B)	0.6	88	—	0.04–6.6
Cadmium (Cd)	1.67–13.1	<1	0.099	0.2–1.5
Chromium (Cr)	<0.003–0.069	0.45	1.99 ± 0.84	0.2–0.4
Cobalt (Co)	0.5–15	1.3–3.0	1.11 ± 0.27	0.1–2.2
Copper (Cu)	1.07–4.19	0.2–1.0	0.21 ± 0.10	6.1–30.3
Gallium (Ga)	0.9 ± 0.03 ng/g	—	—	—
Germanium (Ge)	9 ng/g	—	—	—
Gold (Au)	<0.01–1.1 ng/g	—	0.03–0.01	0.1–83 ng/day
Indium (In)	0.03–0.05	—	—	—
Iron (Fe)	51–148	0.016–1.10	2.18 ± 0.94	0.051–0.350 mg/day
Lead (Pb)	0.10–0.60	6–22	0.46 ± 0.24	8–40
Lithium (Li)	0.01 ± 0.003	—	0.23–3.4	2–10
Manganese (Mn)	0.05–0.85	12–20.2	0.19 ± 0.06	0.08–0.87
Mercury (Hg)	0.054–0.27	1–6	<0.5	1.13–9.8
Molybdenum (Mo)	0.072–0.67	—	—	40–59 μg/day
Nickel (Ni)	0.125	20–83	—	0.4–5.1
Platinum (Pt)	—	—	—	<3
Rubidium (Rb)	2.44–8.40	0.60–0.66	—	1.5
Selenium (Se)	0.79 ± 0.07	11–22	0.28	2–11
Silicon (Si)	23 ± 17	0.342 ± 0.05	—	4700
Silver (Ag)	<5–45 ng/g	0.01	2.18 ± 0.84	0.42–3.8
Strontium (Sr)	0.1 ± 0.02	20	70 ± 18	<0.01–0.03
Tellurium (Te)	<0.05–0.33	—	—	—
Thallium (Tl)	<3 ng/g	—	<40 ng/g	0.1–1.0
Tin (Sn)	0.2 ± 0.04	≤2–3 ng/g	—	0.56–1.6
Vanadium (V)	0.006–0.0033	0.1–0.2	—	<0.2–0.3
Zinc (Zn)	18–47	0.7–4.33	111–227	400–1000

— = no data given.

Source: Tsalev, 1984.

TABLE 1.58 Trace element content of animals

Animal	Body (kg)	Water (g/kg)	Iron	Zinc	Copper	Iodine	Selenium
					mg/kg		
Adult							
Chicken	2.0	760	40	35	1.3	0.4	0.25
Human	65.0	720	74	28	1.7	0.7	0.2
Pig	125.0	750	90	25	2.5	—	0.2
Cat	4.0	740	60	23	1.5	—	0.2
Rabbit	2.6	730	60	50	1.5	—	—
Rat	0.35	720	60	30	2.0	—	—
Steer	500.0	550	168	—	—	0.1	0.1
Newborn							
Chicken	0.04	830	40	—	1.3	—	0.2
Human	3.56	820	94	19.2	4.7	—	—
Pig	1.26	820	29	10.1	3.2	—	0.15
Cat	1.18	822	55	28.7	2.9	—	—
Rabbit	0.54	865	135	22.5	4.0	—	—
Rat	0.006	862	59	24.4	4.3	—	—

Source: Miller et al., 1991.

Much of the interest surrounding the trace elements deals more with the clinical aspects rather than their content per se in various body tissues, for as the concentration changes, the biological effect changes, as shown in Figure 1.5.

Table 1.59 gives the properties associated with several trace elements, which when present in very low concentrations (identified as ultratrace elements) result in a deficiency that can be clinically diagnosed.

The clinical features of some trace elements when deficient and at toxic levels are shown in Table 1.60.

In 1989, the U.S. Food and Drug Administration established Recommended Daily Allowances (RDA) of the essential trace elements chromium, copper, iron, manganese, molybdenum, selenium, and zinc for adults. A listing of the recommended and safe dietary intake for adults for nine elements is given in Table 1.61.

Pais (1995) prepared a comprehensive list of trace elements and the relationship of daily intake to animal and human health; that listing is given in Table 1.62.

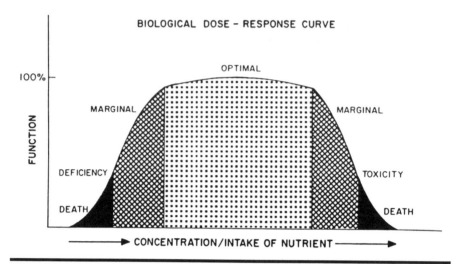

FIGURE 1.5 Dependence of animal function on intake of an essential nutrient according to Mertz (source: McDowell, 1992).

TABLE 1.59   Properties of the essential trace elements

Ultratrace Element	Deficiency Signs	Specific Function
Cadmium (Cd)	Growth depression	Stimulates elongation factors in ribosomes
Chromium (Cr)	Insulin resistance	Potentiation of insulin action on carbohydrates and lipids, active as a bioorganic chromium complex
Cobalt (Co)	Anemia, growth retardation	Constituent of vitamin B_{12}
Lithium (Li)	Growth depression, reduced reproduction	Control of sodium pump
Manganese (Mn)	Growth depression, bone deformities, membrane abnormalities	Carbohydrate metabolism, superoxide dismutase, pyruvate carboxylase, etc.
Nickel (Ni)	Growth depression, reduced N utilization, reduced iron (Fe) metabolism	Constituent of ureases, reduced hemopoiesis
Tin (Sn)	Growth depression	Interactions with riboflavin
Vanadium (V)	Growth depression, inhibition of ATPase, p-transferases	Control of sodium pump

TABLE 1.60   Selected clinical features of trace element deficiencies and toxicities

Trace Element	Deficiency	Toxicity
Chromium (Cr)	Impaired glucose tolerance Fasting hyperglycemia Glycosuria Impaired FFA release Increased serum cholesterol Increased triglycerides Relative insulin resistance Peripheral neuropathy	Trivalent Cr: low toxicity; no well-recognized toxicity syndrome in humans Hexavalent Cr: more toxic; produces elevated insulin or lung tumors in animals
Copper (Cu)	Growth failure (infants) Hair abnormalities Scorbutic bone changes Anemia Neuropenia EKG changes Cardiomegaly and failure	Anemia Diarrhea Abdominal pain Liver and renal disease Coma, death Rare metabolic condition (Wilson's disease)
Fluorine (F)	Increased incidence of dental caries Tooth mottling Deficiency unknown	Fluorosis Neurological disturbances
Iodine (I)	Goiter Cretinism Hypothyroidism	Thyrotoxicosis Goiter
Iron (Fe)	Weakness, fatigue Microcytic hypochromic anemia Reduced resistance to infection	Hemochromatosis Hemosiderosis
Selenium (Se)	Cardiomyopathy Severe muscle pain Keshan disease Kashin-Beck disease	Garlic odor Vomiting Diarrhea Muscle spasm Endemic Se toxicosis
Zinc (Zn)	Alopecia Characteristic dermatitis Anorexia Immune dysfunction Growth retardation Hypogonadism Weight loss	Nausea, vomiting Diarrhea Fever Lethargy Muscle pain Anemia Acute renal failure

TABLE 1.60   Selected clinical features of trace element deficiencies and toxicities (continued)

Trace Element	Deficiency	Toxicity
Zinc (Zn)	Neuropsychiatric symptoms Impaired dark adaptation Rare metabolic condition (acrodermatitis enteropathica)	Pancreatitis Death

Source: Danford, 1989.

TABLE 1.61   Recommended safe and adequate dietary intake for adults (1989)

Trace Element	Intake (mg/day)
Chromium (Cr)	0.05–0.2
Copper (Cu)	2.00–3.0
Fluorine (F)	1.50–4.0
Iodine (I)	0.15
Iron (Fe) (males)	10
Iron (Fe) (females)	18
Manganese (Mn)	2.50–5.0
Molybdenum (Mo)	0.15–0.5
Selenium (Se)	0.02–0.07
Zinc (Zn)	15

TABLE 1.62   Comprehensive listing of daily intake of the trace elements associated with deficiency, optimum level, and beginning of toxicity (elements listed by increasing concentration)

Element	μg/day		
	Deficiency	Optimum	Beginning of Toxicity
Cadmium (Cd)	0.5	1–5	30
Mercury (Hg)	0.5	1–5	50
Thallium (Tl)	0.5	1–5	100
Aluminum (Al)	10	20–100	2,000
Antimony (Sb)	10	50–200	10,000
Arsenic (As)	10	50–100	20,000
Barium (Ba)	10	20–50	1,000

TABLE 1.62 Comprehensive listing of daily intake of the trace elements associated with deficiency, optimum level, and beginning of toxicity (elements listed by increasing concentration) (continued)

Element	µg/day Deficiency	µg/day Optimum	mg/day Beginning of Toxicity
Beryllium (Be)	5	10–20	1
Bismuth (Bi)	10	100–500	10
Chromium (Cr)	20	50–200[a]	5
Cobalt (Co)	10	20–50	500
Gallium (Ga)	50	50–200	5
Germanium (Ge)	1	10–100	1
Gold (Au)	20	50–200	5
Hafnium (Hf)	100	200–600	100
Indium (In)	50	50–200	3
Iodine (I)	10	100–150[a]	5
Lead (Pb)	1	10–20	1
Nickel (Ni)	50	100–200	20
Niobium (Nb)	50	100–500	100
Scandium (Sc)	10	50–100	1
Selenium (Se)	5	20–70[a]	5
Silver (Ag)	20	50–200	5
Strontium (Sr)	50	100–200	50
Tantalum (Ta)	50	100–500	500
Tungsten (W)	10	100–200	50
Tellurium (Te)	10	50–200	5
Vanadium (V)	50	200–400	200
Pt metals	10	20–50	10
Rare earth metals	20	100–200	10

	µg/day	mg/day	mg/day
Boron (B)	100	1–50	500
Fluorine (F)	100	1–3	10
Lithium (Li)	10	20–30	200
Molybdenum (Mo)	50	0.15–0.50[a]	5
Tin (Sn)	10	2–10	2,000
Copper (Cu)	1,000	2–3[a]	200
Iron (Fe)	1,000	10–20[a]	200
Manganese (Mn)	1,000	3–5[a]	40
Zinc (Zn)	1,000	10–15[a]	600
Cesium (Cs)	100	5–10	100
Silicon (Si)	5,000	50–100	500
Rubidium (Rb)	100	5–10	100

TABLE 1.62   Comprehensive listing of daily intake of the trace elements associated with deficiency, optimum level, and beginning of toxicity (elements listed by increasing concentration) (continued)

Element	Deficiency	Optimum	Beginning of Toxicity
	mg/day	mg/day	g/day
Bromine (Br)	1	10–30	1
Titanium (Ti)	1	20–100	10
Zirconium (Zr)	1	10–50	1

Note: Favorable or unfavorable role of the given element depends on the compound form or the oxidation state of an inorganic salt, methylated compound, such as Cr(III) or Cr(VI), etc., between different elements. There are complicated interactions; therefore, for a realistic evaluation, the existence of strong synergistic or antagonistic effects among the elements should be known. With any given element, there can be considerable differences in levels of response depending on the age, sex, etc. of the individual. With regard to animals, sensitivity responses will vary, which means that what might be toxic for horses might not be so for chickens, etc.

[a] These values are the internationally accepted RDA values (1989).

Source: Pais, 1995.

Along the same lines as Pais (1995), McDowell (1992) prepared what would be the maximum tolerable levels for the 26 trace elements on 6 animal species (Table 1.63).

In the previous sections, trace element concentrations in various parts of the environment have been presented, drawing on the wealth of information found in the literature. Pais (1991) summarized the average expected con-

TABLE 1.63   Maximum tolerable levels of dietary minerals for domestic animals[a]

Trace Element	Species					
	Cattle	Sheep	Swine	Poultry	Horse	Rabbit
Aluminum (Al)[b]	1000	1000	(200)	200	(200)	(200)
Antimony (Sb)	—	—	—	—	—	70–150
Arsenic (As)						
Inorganic	50	50	50	50	(50)	50
Organic	100	100	100	100	(100)	(100)
Barium (Ba)[b]	(20)	(20)	(20)	(20)	(20)	(20)

TABLE 1.63   Maximum tolerable levels of dietary minerals for domestic animals[a] (continued)

Trace Element	Cattle	Sheep	Swine	Poultry	Horse	Rabbit
			Species			
Bismuth (Bi)	(400)	(400)	(400)	(400)	(400)	2000
Boron (B)	150	(150)	(150)	(150)	(150)	(150)
Bromine (Br)	200	(200)	200	2500	(200)	(200)
Cadmium (Cd)[c]	0.5	0.5	0.5	0.5	(0.5)	(0.5)
Chromium (Cr)						
Chloride	(1000)	(1000)	(1000)	1000	(1000)	(1000)
Oxide	(3000)	(3000)	(3000)	3000	(3000)	(3000)
Cobalt (Co)	10	10	10	10	(10)	(10)
Copper (Cu)	100	25	250	300	800	200
Fluorine (F)[d]	Young, 40 Mature dairy, 40 Mature beef, 50 Finishing, 100	Breeding, 60 Finishing, 150	150	Turkey, 150 Chicken, 200	(40)	(40)
Iodine (I)	50[e]	50	400	300	5	—
Iron (Fe)	1000	500	3000	1000	(500)	(500)
Lead (Pb)[c]	30	30	30	30	30	(30)
Manganese (Mn)	1000	1000	400	2000	(400)	(400)
Mercury (Hg)[c]	2	2	2	2	(2)	(2)
Molybdenum (Mo)	10	10	20	100	(5)	500
Nickel (Ni)	50	(50)	(100)	(300)	(50)	(50)
Selenium (Se)	(2)	(2)	2	2	(2)	(2)
Silicon (Si)[b]	(11.2)	—	—	(0.2)	—	—
Silver (Ag)	—	(100)	100	—	—	—
Strontium (Sr)	2000	(2000)	3000	Laying hen, 30,000 Other, 3000	(2000)	(2000)
Tungsten (W)	(20)	(20)	(20)	20	(20)	(20)
Vanadium (V)	50	50	(10)	10	(10)	(10)
Zinc (Zn)	500	300	1000	1000	(500)	(500)

[a] Taken from NRC. 1980. Mineral Tolerance of Domestic Animals. National Academy of Sciences, National Research Council, Washington, D.C. The accompanying text should be consulted prior to applying the maximum tolerable levels in practical situations. Continuous long-term feeding of minerals at the maximum tolerable levels may cause adverse effects. The listed levels were derived from toxicity data on the designated species. The levels in parentheses were derived by interspecific extrapolation. Dashes indicate that data were insufficient to set a maximum tolerable level.

TABLE 1.63   Maximum tolerable levels of dietary minerals for domestic animals[a] (continued)

[b] As soluble salts of high bioavailability. Higher levels of less soluble forms found in natural substances can be tolerated.

[c] Levels based on human food residue considerations.

[d] As sodium fluoride or fluorides of similar toxicity. Fluoride in certain phosphate sources may be less toxic. Morphological lesions in cattle teeth may be seen when dietary fluoride for the young exceeds 20 ppm, but a relationship between the lesions caused by fluoride levels below the maximum tolerable levels and animal performance has not been established.

[e] May result in undesirably high iodine levels in milk.

Source: McDowell, 1992.

centration of the trace elements in various components in the environment, as shown in Table 1.64.

FOOD/FEED AND FOOD PRODUCTS

In the United States, the trace element content of food plants became a major issue as assays of farm and food products were being made routinely as a result of improved analytical procedures that allowed the analyst to easily assay large numbers of samples at low cost (see Chapter 6). Also of interest was the so-called "background" level of the trace elements existing in common raw agricultural products that would not be affected by human activity. Such background levels would be difficult to obtain in much of the world due to centuries of human activity. In a joint project between several government agencies, raw agricultural crops were collected from carefully selected areas within the United States and assayed; the analytical data were then published between 1983 and 1985.

In the initial publication, cadmium and lead were the two elements reported, with nickel, copper, iron, manganese, molybdenum, selenium, and zinc assay data coming in later publications. The cadmium, lead, and nickel data for 12 raw agricultural products are given in Table 1.65, and the copper, iron, manganese, molybdenum, selenium, and zinc data are given in Table 1.66.

In the United States, several total diet studies have been initiated to determine the microelement (copper, iodine, iron, manganese, selenium, and zinc) contents in 234 commonly consumed foods and food products. The copper, iron, manganese, and zinc contents of many foods and food products had been published earlier by Shacklette (1980) and Gebhardt et

TABLE 1.64   The average expected concentration of trace elements

Element	Earth (weight-ratio)	Sea water (mg/L)	mg/kg Plants	mg/kg Animals	mg/kg Humans
Aluminum (Al)	81,300	0.01	500	4–100	<0.8
Boron (B)	3.0	4.6	50.0	0.5	<1.0
Cobalt (Co)	23	0.00027	0.5	0.03	0.02
Copper (Cu)	70	0.003	14.0	2.4	1.6
Chlorine (Cl)	314	19,000	2,000	2,800	800
Chromium (Cr)	200	0.00005	0.2	0.07	0.07
Fluorine (F)	300	1.3	1–40	350–800	600
Iodine (I)	0.3	0.06	0.42	0.4	0.2
Iron (Fe)	50,000	0.01	140	160	100
Lithium (Li)	65	0.18	0.1	0.02	0.02
Manganese (Mn)	1,000	0.002	120	0.2	0.3
Molybdenum (Mo)	15	0.01	0.9	0.2	0.2
Nickel (Ni)	80	0.0054	3.0	0.8	0.15
Selenium (Se)	0.09	0.00009	0.2	1.7	0.2
Titanium (Ti)	4,400	0.001	1.0	0.2	<0.02
Vanadium (V)	150	0.002	1.6	0.1	0.2
Zinc (Zn)	132	0.01	100	0.3	30

Source: Pais, 1991.

TABLE 1.65   Cadmium, lead, and nickel content of raw agricultural crops in the United States

Crop	μg/g wet weight Cadmium	μg/g wet weight Lead	μg/g wet weight Nickel
Carrots	0.028	0.009	0.071
Field corn	0.012	0.022	0.33
Lettuce	0.026	0.013	—
Onion	0.011	0.005	0.36
Peanut	0.078	0.100	2.0
Potato	0.031	0.009	—
Rice	0.012	0.007	0.285
Soybean	0.059	0.042	4.8
Spinach	0.065	0.045	0.098
Sweet corn	0.0031	0.0033	0.062
Tomato	0.017	0.002	0.070
Wheat	0.043	0.037	—

Source: Wolnik et al., 1983a, b.

TABLE 1.66 Copper, iron, manganese, molybdenum, selenium, and zinc content of raw agricultural crops in the United States

			μg/g wet weight			
Crop	Copper (Cu)	Iron (Fe)	Manganese (Mn)	Molybdenum (Mo)	Selenium (Se)	Zinc (Zn)
Carrots	0.58	3.2	1.5	0.015	—	2.6
Field corn	1.5	18.0	5.1	0.22	—	18.5
Lettuce	0.26	3.0	1.8	0.013	0.0016	1.9
Onion	6.37	1.45	1.2	0.016	—	1.7
Peanut	7.6	18.0	16.0	0.043	0.057	28.0
Potato	0.96	3.85	1.4	0.036	0.0030	3.1
Rice	1.9	3.6	12.0	0.66	—	13.5
Soybean	12.0	66.0	27.0	2.1	0.19	42.0
Spinach	0.66	17.0	7.1	0.024	—	4.5
Sweet corn	0.45	4.0	1.6	0.048	0.0064	5.6
Tomato	0.645	3.0	1.1	0.024	—	1.4
Wheat	4.4	32.5	38.0	0.40	0.37	27.0

Source: Wolnik et al., 1983b, 1985.

al. (1982) and more recently for the trace elements copper, iodine, iron, manganese, selenium, and zinc for core foods in the U.S. food supply by Pennington et al. (1995). However, with markedly improved analytical procedures, requirements for food labeling, and concerns regarding the presence of particular elements in some food products that could be at potentially toxic concentrations, total diet studies were begun and are still ongoing, assaying food and food products taken directly from food markets and stores.

Based on analytical data gathered from 1982 to 1989, the results of a total diet study were published by Pennington and Young (1990), who summarized the average concentration of microelements in foods by food group on a per 100 g and per typical serving portion basis. The average concentrations of six microelements (copper, iodine, iron, manganese, selenium, and zinc) in food by food product on a milligram per serving basis are shown in Table 1.67.

Pennington and Young (1990) summarized their findings as follows:

> Foods highest per serving in Cu were meat, nuts, mixed dishes, and beans/peas; for I, ready-to-eat cereals, dairy desserts, mixed dishes, fish, and dairy products; for Fe, ready-to-eat cereals, mixed

TABLE 1.67   Average concentration of microelements in foods by food group

Food Group (n)[a]	*mg per serving*					
	Copper (Cu)	Iodine (I)	Iron (Fe)	Manganese (Mn)	Selenium (Se)	Zinc (Zn)
Vegetables						
Leafy (7)	0.036	0.002	1.39	0.321	0.000	0.26
Stem/flower (4)	0.045	0.000	0.40	0.135	0.000	0.22
Beans/peas (9)	0.164	0.018	1.72	0.430	0.001	0.94
Root/tuber (13)	0.087	0.016	0.61	0.225	0.000	0.29
Other (17)	0.072	0.003	0.54	0.124	0.000	0.30
Fruit						
Fruit juices (25)	0.071	0.004	0.42	0.234	0.000	0.11
Grain products						
Cooked grains (5)	0.051	0.027	1.87	0.375	0.005	0.44
Ready-to-eat cereals (7)	0.121	0.087	5.79	0.840	0.004	1.30
Bread, rolls, pasta, etc. (13)	0.042	0.020	0.09	0.193	0.007	0.32
Nuts (3)	0.227	0.002	0.62	0.800	0.001	1.14
Eggs (3)	0.031	0.027	1.04	0.015	0.013	0.76
Dairy products (11)	0.017	0.049	0.14	0.017	0.002	1.05
Animal flesh						
Fish (4)	0.027	0.057	0.84	0.079	0.037	0.70
Poultry (3)	0.049	0.015	0.76	0.019	0.022	1.45
Meat (11)	0.625	0.016	2.14	0.045	0.023	4.14
Breakfast/luncheon meat (5)	0.023	0.006	0.50	0.017	0.004	0.83
Mixed dishes (12)	0.185	0.064	2.71	0.382	0.022	2.52
Soups (4)	0.049	0.013	0.80	0.119	0.001	1.17
Desserts						
Dairy-based (5)	0.116	0.070	0.81	0.116	0.000	0.73
Grain-based (7)	0.078	0.024	1.13	0.177	0.004	0.34
Other (5)	0.051	0.027	0.45	0.131	0.000	0.28
Sweeteners (6)	0.034	0.005	0.17	0.042	0.000	0.07
Fats and sauces (9)	0.001	0.001	0.02	0.002	0.000	0.02
Beverages (13)	0.007	0.003	0.12	0.079	0.000	0.02

TABLE 1.67   **Average concentration of microelements in foods by food group (continued)**

Food Group (n)[a]	Copper (Cu)	Iodine (I)	Iron (Fe)	Manganese (Mn)	Selenium (Se)	Zinc (Zn)
			mg per serving			
Strained/junior foods						
Vegetables (7)	0.080	0.003	0.31	0.198	0.000	0.10
Fruits, fruit juices (9)	0.080	0.003	0.31	0.198	0.000	0.10
Cereals (2)	0.071	0.004	9.34	0.527	0.001	0.40
Infant formula (2)	0.025	0.003	0.27	0.005	0.000	0.23
Meat/poultry (3)	0.051	0.017	1.30	0.009	0.010	2.48
Dinners (9)	0.081	0.017	0.86	0.238	0.003	1.00
Dessert (1)	0.034	0.019	0.39	0.061	0.000	0.49

[a] Refers to the number of samples and/or products included in that group.

Source: Pennington and Young, 1990.

dishes, and meats; for Mn, ready-to-eat cereals, nuts, and beans/peas; for Se, fish, meat, poultry, and mixed dishes; and for Zn, meat, mixed dishes, and ready-to-eat cereals. Coefficients of variation for the microelements in the top 20 food sources per serving averaged: 26% for Cu, 104% for I, 28% for Fe, 25% for Mn, 32% for Se, and 20% for Zn.

In a similar summary, Miller et al. (1991) also determined the trace element (iron, zinc, manganese, copper, iodine, cobalt, and selenium) content of common feedstuffs; their results are given in Table 1.68.

Faelten (1981) has graded over 200 foods by the amount of trace elements (zinc, copper, iron, selenium, chromium, manganese, and iodine) they contribute to the diet. An important trace element related to human health is selenium (Oldfield, 1994); therefore, its content in common foods would be of considerable interest (Gebhardt and Holden, 1992). In 1990, the USDA–Human Nutritional Information Service published provisional selenium values for various foods; a partial listing is given in Table 1.69.

The trace element content of commonly consumed foods worldwide would be of more importance than those less commonly consumed. Rice and fish are major foods in the diet of much of the world population; their trace element contents are given in Table 1.70.

A major primary food source for young mammals is milk; therefore, its

TABLE 1.68 Trace element concentrations of common feedstuffs

Feedstuffs	Fe	Zn	Mn	Cu	I	Co	Se
				mg/kg			
Alfalfa meal, 17% protein	200	35	43	10	0.5	0.18	0.05–0.45
Barley	50	17	16	8	0.05	0.1	0.1–0.3
Beet pulp	300	1	35	12.5	—	0.1	—
Blood meal (flash dried)	3800	—	5	9	—	0.1	0.07
Bone meal (steamed)	800	425	30	16	—	0.1	—
Citrus pulp, dried	200	14	7	6	—	—	—
Coconut oil meal	680	—	55	19	—	0.2	
Corn, dent, yellow	35	10	5	4.5	0.05	0.1	0.03–0.38
Corn gluten feed	500	—	25	50	—	0.1	0.2
Corn gluten meal	400	—	7	30		0.1	1.15
Cottonseed meal	100	—	20	20	0.12	0.1	0.06
Distillers dried corn grains	200	—	19	45	—	0.1	—
Distillers dried corn solubles	600	85	74	80	—	—	0.50
Fish meal, herring	300	110	10	20	1.0	—	1.5–2.45
Fish meal, menhaden	270	150	36	8	—	—	1.7
Fish solubles (30%)	300	38	25	48	—	—	1.0
Hominy feed	10	—	14	2	—	0.1	0.1
Linseed meal	300	—	37	25	0.07	0.2	1.1
Meat and bone meal, 50% protein	500	100	19	12	1.3	0.1	0.1–0.8
Milk, cow's	2	4	0.06	0.3	0.04	—	0.04
Molasses, beet	100	—	5	18	1.6	0.4	—
Molasses, cane	100	—	42	60	1.6	0.9	—
Oat grain	70	—	38	6	0.06	0.06	0.05–0.02
Oystershell, ground	2900	—	130	—	—	—	0.01
Peanut meal	20	20	24	30	—	—	0.28
Rice bran	190	30	200	13	—	—	—
Rye grain	45	35	35	6	0.05	—	0.2
Sesame meal	—	100	48	—	—	—	—
Skim milk, dried	8	40	2	3	—	—	0.08–0.15
Sorghum grain	50	17	13	14	0.02	0.1	—
Soybean meal, 44% protein	150	27	35	20	0.13	0.1	0.05–0.10
Soybean meal, dehulled	150	45	40	20	0.1	0.1	0.05–0.10
Wheat bran	150	80	115	12	0.07	0.1	0.6
Wheat grain	50	5	20	7	0.04	0.08	0.05–0.8
Wheat standard middlings	100	150	118	22	0.1	0.1	0.28–0.88
Whey, dried	7	3	5	45	—	0.1	0.008
Yeast, dried, brewer's	50	40	6	30	0.01	0.2	0.11–1.1
Yeast, dried, torula	90	100	13	13	—	0.04	0.03–0.05

Source: Miller et al., 1991.

TABLE 1.69   Selenium content of foods (provisional)

Food	Selenium (µg/100 g edible portion)
Beef products	12.7–32.9
Beverages	0–12.6
Brazil nut	2960
Bread	5.6–48.0
Breakfast cereals	3.1–123.1
Cereal grains/pasta	2.3–79.2
Crackers	4.8–34.8
Dairy products	0.6–26.2
Eggs	17.6–45.2
Fast food	16.3–21.4
Fats and oils	0.2–2.1
Fish and shellfish	12.6–80.4
Fruits/fruit juices	0.1–1.0
Legumes	0.8–18.5
Mustard	36.0
Nuts/seeds	4.6–78.2
Pork products (liver: 190)	8.0–51.6
Poultry	21.7–70.9
Sausages/luncheon meat	11.3–58.0
Snacks/sweets	0.6–17.8
Soups/sauces/gravies	0.4–13.6
Vegetables	0.4–14.2

Source: USDA–Human Nutrition Information Service, HN1S/PT-109.

trace element composition would be important for the health and well-being of newborns and the young. The trace element content of human milk and three other mammals is given in Table 1.71.

The role that pollution may play in the trace element content of foods and food products was illustrated in a study conducted by Fodor and Molnár (1993), who collected honey from known unpolluted and polluted sites at three locations in Hungary (Table 1.72). The comparison of trace element contents between the two pairs of honey samples clearly shows the influence that pollution can have on a food product that one would not expect to be so easily affected.

Up to the present, the primary focus of trace element content has been on what might be considered as high rather than low levels in foods and

TABLE 1.70 Trace element content of some commonly consumed items in the diet for a large portion of the world population

Element	*mg/kg*		
	Rice Flour	*Copepoda*	*Fish Flesh*
Arsenic (As)	0.43	7.0	2.8
Bromine (Br)	0.98	950.0	21.9
Cobalt (Co)	0.02	0.13	0.08
Copper (Cu)	1.95	7.5	3.6
Iron (Fe)	9.0	61.3	52.2
Manganese (Mn)	18.9	3.0	0.92
Mercury (Hg)	0.008	0.31	0.55
Rubidium (Rb)	7.8	3.5	8.2
Selenium (Se)	0.48	2.8	1.9
Zinc (Zn)	22.0	166.0	34.0

Source: Toro et al., 1994.

food products. There is still much to be learned regarding the background level of many of the trace elements found in food plants, foods, and food products (see Tables 1.65 and 1.66). Therefore, what might be considered high or low could be the naturally occurring background level.

TABLE 1.71 Trace element content in milk by animal species

Element	*mg/L*			
	Cow	*Horse*	*Guinea Pig*	*Human*
Aluminum (Al)	0.10	0.12	0.45	0.12
Barium (Ba)	0.19	0.08	0.22	0.15
Boron (B)	0.33	0.10	0.59	0.27
Copper (Cu)	0.05	0.15	0.51	0.31
Iron (Fe)	0.19	0.22	0.72	0.26
Lithium (Li)	0.02	0.01	0.03	0.01
Manganese (Mn)	0.02	0.01	0.01	0.01
Molybdenum (Mo)	0.02	0.01	0.02	0.02
Silicon (Si)	0.43	0.16	0.58	0.47
Strontium (Sr)	0.42	0.44	1.09	0.06
Titanium (Ti)	0.11	0.14	trace	0.02
Zinc (Zn)	3.96	1.83	4.23	2.15

Source: Anderson, 1992.

TABLE 1.72   Trace element content of some Hungarian honeys

| | mg/kg | | | | | |
| | Non-Polluted Area | | | Polluted Area | | |
Element	Acacia	Sunflower	Lindenfl.	Acacia	Sunflower	Lindenfl.
Aluminum (Al)	4.2	2.1	38.0	6.2	7.0	45.6
Boron (B)	6.0	0.6	0.2	5.2	0.6	0.2
Cadmium (Cd)	0.06	0.03	0.06	0.1	0.1	0.12
Copper (Cu)	3.6	0.2	1.7	2.7	0.6	3.4
Iron (Fe)	6.5	1.4	36.0	24.0	56.0	9.7
Lead (Pb)	1.4	0.6	0.7	1.8	2.4	1.9
Manganese (Mn)	0.6	1.5	5.8	0.8	7.1	1.8
Zinc (Zn)	14.0	12.0	48.0	18.0	84.0	36.0

Source: Fodor and Molnár, 1993.

TABLE 1.73   Ranges and medium concentrations of trace elements in dry digested sewage sludges

| | Reported Range (mg/kg) | | |
Element	Minimum	Maximum	Median
Arsenic (As)	1.1	230	10
Cadmium (Cd)	1.0	3,400	10
Chromium (Cr)	10	99,000	500
Cobalt (Co)	11.3	2,490	30
Copper (Cu)	84	17,000	800
Fluorine (F)	80	33,500	260
Iron (Fe)	1,000	154,000	17,000
Lead (Pb)	13	26,000	500
Manganese (Mn)	32	9,870	260
Mercury (Hg)	0.6	56	6
Molybdenum (Mo)	0.1	214	4
Nickel (Ni)	2	5,300	80
Tin (Sn)	2.6	329	14
Selenium (Se)	1.7	17.2	5
Zinc (Zn)	101	49,000	1,700

Source: Chaney, 1983.

In addition, the long-term effects of man's activity are now just becoming better understood, with concerns primarily focused on the effects of land disposal of waste products, such as sewage sludge (which contains a wide range of trace element contents, as shown in Table 1.73), and on plant growth, plant composition, and raw food products (Page et al., 1987).

One mechanism of transfer of the trace elements found in sewage sludge land applied into the food chain would be from the soil into forage and into grazing domestic animals, as can be seen from the data given in Table 1.74.

The introduction of the trace elements into the food chain from the soil into commonly consumed food plants was based partly on the ability of the plant to accumulate a particular trace element (Table 1.75).

Some trace elements are readily taken up by plants and transported, while others are not (see Tables 1.43, 1.52, and 1.53). Since most of the

TABLE 1.74 Maximum tolerable levels of dietary minerals for domestic livestock in comparison with levels in forages

Element	Soil–Plant Barrier	Level in Plant Foliage[a] (mg/kg dry foliage)		Maximum Levels Chronically Tolerated[b] (mg/kg dry diet)			
		Normal	Phytotoxic	Cattle	Sheep	Swine	Chicken
As	yes	0.01–1	3–10	50	50	50	50
B	yes	7–75	75	150	(150)	(150)	(150)
Cd	fails	0.1–1	5–700	0.5	0.5	0.5	0.5
Cr^{3+}	yes	0.1–1	20	(3000)	(3000)	(3000)	3000
Co	fail?	0.01–0.3	25–100	10	10	10	10
Cu	yes	3–20	25–40	100	25	250	300
F	yes?	1–5	—	40	60	150	200
Fe	yes	30–300	—	1000	500	3000	1000
Mn	?	15–150	400–2000	1000	1000	400	2000
Mo	fail?	0.1–0.3	100	10	10	20	100
Ni	yes	0.1–5	50–100	50	(50)	(100)	(300)
Pb	yes	2–5	—	30	30	30	30
Se	fails	0.1–2	100	(2)	(2)	2	2
V	yes?	0.1–1	10	50	50	2	10
Zn	yes	15–150	500–1500	500	300	1000	1000

[a] Based on NRC, 1980b. Continuous long-term feeding of minerals at the maximum tolerable levels may cause adverse effects. Levels in parentheses were derived by interspecific extrapolation by NRC.
[b] Maximum levels tolerated based on human food residue consideration.

Source: Chaney, 1983.

TABLE 1.75   Relative accumulation of cadmium and lead in edible portions and copper, nickel, and zinc in leaves

Element	High Accumulators	Low Accumulators
Cadmium (Cd)	Lettuce, spinach, celery, cabbage	Potato, maize, french bean, peas
Copper (Cu)	Sugar beet, certain barley cultivars	Leek, cabbage, onion
Lead (Pb)	Kale, ryegrass, celery	Some barley cultivars, potato, maize
Nickel (Ni)	Sugar beet, ryegrass, marigold	Maize, leek, barley cultivars, onion
Zinc (Zn)	Sugar beet, marigold, spinach, red beet	Potato, leek, tomato, onion

determined trace elements are considered essential for normal health and growth for both animals and humans, diets consisting primarily of those feeds/foods low in these elements might lead to possible health problems unless such elements are brought into the body by some other ingested substances (Hamilton and Minski, 1973). Therefore, selecting feeds/foods and food products that have a mix of high and low contents into the daily diet would seem nutritionally wise. The effects of dietary deficiencies of trace elements in the diet are just now being discovered and better understood. Unfortunately, much of the emphasis on the trace element content in common feeds/foods has been on those elements that are high, when low levels would be equally or more important in terms of their impact on animal and human health.

Although a particular feed/food or food product may contain a substantial quantity of a trace element, adsorption and utilization by the animal or human body may not occur due to other factors. However, this aspect of elemental utilization is beyond the scope of this book and readers should refer to other sources for such information.

2

TRACE
ELEMENTS

In this chapter, 41 elements are described in terms of their physicochemical properties, mineral source(s), and concentrations found in soil, water, the human body, animals, plants, fertilizers, and foods. The objective is to bring together in an easily useable form as much information about a trace element as is known and available in the literature.

The information and concentration values given in this chapter were taken from a number of sources, mostly from those references listed in this book. Many of the values given are not consistently quoted in the current literature; therefore, the reader may find discrepancies. However, the authors have attempted to use only those values given in authoritative sources.

For most of the trace elements, there is a wealth of information available; for some trace elements, little data was found. This chapter does not, therefore, provide an exhaustive listing of known values, but rather the information has been taken from readily available reference books, manuals, and scientific articles and then condensed for ease of use.

Our goal is to provide, in one source, as much descriptive information as is available for each trace element related to various biological substances and systems.

ALUMINUM (Al)

Soft and malleable metal protected by oxide film from reacting with air and water; soluble in hot concentrated HCl and NaOH solution

Atomic Number: 13   **Abundance in Lithosphere**: 83,000 mg/kg
Atomic Weight: 26.9815   **Common Valence State**: Al^{3+}

Common Mineral Forms: bauxite, found as beohmite [AlO(OH)] and gibbsite [Al(OH)$_3$]

Total Content in Soils: 0.45–10%

Soluble Content in Soils: 400 µg/L (saturated paste)

Content in Sea Water: 0.13×10^{-4} to 9.7×10^{-4} mg/L

Content in Fresh Water: 0.1–1200 µg/L (reference 200 µg/L)

Chemical Species in Water: $Al(OH)_4^-$, possibly: Al^{3+}, $AlOH^{2+}$, $Al(OH)_3$

Content in Humans: muscle, 0.7–28 mg/kg; bone, 4–27 mg/kg; blood, 0.39 mg/dm^3; milk, 0.12–0.33 mg/L; kidney, 0.4 µg/g; tooth, 8–325 µg/g; urine: 3.5–31 µg/L; scalp hair, 0.813 mg/kg

Content in Animals: 4–100 mg/kg; cow and horse milk, 0.12 mg/L

Content in Plants: 10–1000 mg/kg (average 200 mg/kg); reference plant, 80 mg/kg

Essentiality: plants, no; animals, no

General

Aluminum is the third most abundant element in the lithosphere and the most abundant of the trace elements. Most naturally occurring aluminum compounds are insoluble; therefore, relatively small quantities of aluminum are found in most biological systems unless contaminated with soil (dust).

Soils/Plants

Aluminum is toxic to plants in concentrations from 0.1 to 3.0 mg/L in nutrient and/or soil solution. With increasing soil acidity, high aluminum availability may be the cause of the decline in native forests as well as reduced crop growth in cultivated soils. Aluminum soil availability can be easily controlled by liming an acid (pH <6.0) soil and maintaining the soil water pH between 6.0 and 7.0. Aluminum toxicity in plants is essentially indirect, affecting root growth and the uptake of essential elements, particularly phosphorus, and calcium-magnesium antagonism, while aluminum toxicity is associated with these proposed effects. High aluminum levels (>100 mg/kg) in stems and leaves of plants may occur when plant roots are exposed to anaerobic conditions. The aluminum level in young plant seedlings may be very high (>500 mg/kg) without detrimental effect, with a significant rapid decline in aluminum concentration as the plant begins its juvenile growth period. An accurate determination of aluminum in plant tissue can only be made if the tissue is free of soil and/or dust particles.

Animals/Man

Daily dietary intake is 2.45 mg, toxic intake is 5 g, and total mass of the element in an average (70-kg) person is 61 mg. Aluminum is implicated in Alzheimer's disease (senile dementia). Intake of 200 mg Al^{3+} in a 10-g/day dry weight diet (20,000 ppm) is toxic to rats, whereas 220 mg (22,000 ppm) is lethal. Chronic toxicity results in several phosphorus metabolism imbalances, including excretion of phosphorus, decreased incorporation into phospholipids, a drop in ATP, and a rise in ADP levels in the blood.

Mobility in Food Chain

By calcium channels with the help of transferrin.

ANTIMONY (Sb)

Metalloid element with various allotropes, of which metal is bright, silvery, hard, and brittle; stable in dry air and not attacked by dilute acid or alkalis

Atomic Number: 51   **Abundance in Lithosphere**: 0.2 mg/kg
Atomic Weight: 121.75   **Common Valence States**: Sb^{3+}, Sb^{5+}

Common Mineral Forms: stibnite (Sb_2S_3) and ullmanite (NiSbS)

Total Content in Soils: mean 0.9 mg/kg; range 0.01–1.0 mg/kg

Soluble Content in Soils: 0.5 mg/kg

Content in Sea Water: 2.4×10^{-4} mg/kg

Content in Fresh Water: reference level 0.2 μg/L

Chemical Species in Water: $Sb(OH)_6^-$

Content in Humans: muscle, 0.042–0.191 mg/kg; bone, 0.01–0.6 mg/kg; blood, 0.0033 mg/dm^3

Maximum Daily Intake from Toy Materials: 0.2 μg

Content in Animals: 40–400 mg/kg in liver; 400 mg/kg considered lethal

Content in Plants: 0.1–200 μg/kg; reference plant, 0.1 mg/kg

Content in Common Foods: 0.02–4.3 μg/kg

Essentiality: plants, no; animals, no

General

Although antimony primarily exists in cationic forms, it also has an anionic form, such as antimonides and antimonates.

Plants

Little is known about antimony and its effect on plants, although it is thought to be moderately toxic to plants. It has been found that plants will accumulate antimony above that found in their rooting media; 7 to 50 mg/kg has been found in trees and shrubs growing in high antimony-containing soils. Normal antimony ranges in leaves would be 10 to 27 μg/kg.

Animals/Man

Toxic intake is 100 mg. Sb(V) forms are more toxic than Sb(III) forms. Ingested Sb(V) does not penetrate cell walls and is excreted through urine. Some antimony compounds are used as treatment for several diseases (leishmaniasis).

Mobility in Food Chain

Antimony has a relatively high mobility in the environment and has a moderate bioaccumulation index.

ARSENIC (As)

Metalloid with several allotropes; grey α-arsenic is metallic—brittle; tarnishes; burns in oxygen; resists attack by water, acids, and alkalis; attacked by hot acids and molten NaOH

Atomic Number: 33	**Abundance in Lithosphere**: 1.5 mg/kg
Atomic Weight: 74.9216	**Common Valence State**: As^{5+} and As^{3+}

Common Mineral Forms: realgar (As_4S_4), orpiment (As_2S_3), loellingite ($FeAs_2$)

Total Content in Soils: 0.1–48 mg/kg; mean 3.6–8.8 mg/kg

Soluble Content in Soils: using P-soil test extractants, about 40% of total

Content in Sea Water: 1.45×10^{-3} mg/kg

Content in Fresh Water: 0.1–800 ng/g; reference value 0.5 µg/L

Chemical Species in Water: $HAsO_4^{2-}$, $H_2AsO_4^-$, possibly H_3AsO_4, AsO_4^{3-}, or $H_2AsO_3^-$

Content in Animals: mean range 0.04–0.09 µg/g dry weight, higher in skin, nails, and hair

Content in Humans: muscle, 0.009–0.65 mg/kg; bone, 0.08–1.6 mg/kg; blood, 0.0017–0.09 mg/dm^3; liver, 1–13 ng/g; milk, 1.6–6.0 mg/L

Maximum Daily Intake from Toy Materials: 0.1 µg

Content in Plants: 0.009–1.7 mg/kg (toxic when 5–10 mg/kg); reference plant, 0.1 mg/kg

Content in Fertilizers: 0.4–188 mg/kg

Content in Common Foods: 0.21–16 mg/kg; U.S. Public Health Service maximum for fruits and vegetables, 2.6 mg/kg fresh weight

Essentiality: plants, no; animals, yes

General

In the past, arsenic-containing compounds had been widely used as pesticides, herbicides, and soil sterilants and were found in fairly high concentrations (0.2%) in some agricultural soils. In addition, arsenic contamination has occurred as a result of some industrial activities (e.g., emissions from coal-fired power plants). The organic forms of arsenic, such as the methyl-, dimethyl-, and trimethyl-arsenic acid forms, are more toxic than the elemental forms. Arsenic reaction and movement in soils are dependent on its oxidation state. Deep waters off of Taiwan and waters of southeast Hungary have been found to contain toxic concentrations of arsenic.

Plants

Arsenic has been found to be phytotoxic to tomato plants at concentrations greater than 2 mg/L in nutrient solution (range 0.02 to 7.5 mg/L), with arsenic essentially accumulating in the roots with relatively slow and minimal translocation from the root. Arsenic greater than 2 mg/kg dry weight in plants is generally phytotoxic. The phytotoxicity of arsenic is reduced with high phosphorus availability. The level of arsenic in plants increases with the increasing level in the rooting medium; the degree of increase varies with crop species.

Animals/Man

Daily dietary intake is 0.04 to 1.4 mg (daily required intake for adult humans is 12 to 25 µg), toxic intake is 5 to 50 mg/day, and lethal intake is 50 to 340 mg/day. Arsenic is believed to be carcinogenic. In most human biological materials, the concentration of arsenic is well below 1 µg/g. Animals become arsenic deficient when rations contain less than 50 ng/g, while the normal supply level should be between 350 to 500 ng/g.

Mobility in Food Chain

Arsenic is not highly mobile and therefore would not be expected to accumulate from movement within the food chain. The major source of arsenic would be from physical deposition or from consuming products high in arsenic, such as seafood. The bioaccumulation index for arsenic is moderate.

BARIUM (Ba)

Relatively soft, silvery-white metal; attacked by air and water

Atomic Number: 56   **Abundance in Lithosphere**: 425 mg/kg
Atomic Weight: 137.36   **Common Valence State**: Ba^{2+}

Common Mineral Forms: barite ($BaSO_4$), witherite ($BaCO_3$), beryllite ($Be_2BaSi_2O_7$); quartz-diorite contained in higher amounts; barium in magmatic rock ranges from 0.5 to 1200 mg/kg and in sedimentary rock from 50 to 800 mg/kg

Total Content in Soils: 100–3000 mg/kg; mean value 500 mg/kg

Content in Sea Water: 13 µg/L

Content in Fresh Water: 2.6 µg/L; reference value 10 µg/L

Chemical Species in Water: Ba^{2+}

Content in Marine Plants: 30 mg/kg (accumulates in brown algae)

Content in Humans: muscle, 0.09 mg/kg; bone, 3–70 mg/kg; blood, 0.068 mg/dm^3; urine, 0.005 mg/L; scalp hair, 1.09 mg/kg; milk, 20 mg/L

Maximum Daily Intake from Toy Materials: 25.0 µg

Content in Animals: 0.8 mg/kg (accumulates in bones in high concentrations)

Content in Plants: 15 mg/kg (can accumulate in nut tree leaves in excess of 10,000 mg/kg), range 1–198 mg/kg; reference plant, 40 mg/kg

Content in Common Foods: less than 0.5 mg/kg, but Brazil nuts may contain 3 mg or higher amounts

Content in Fertilizers: 100 mg/kg (range 1–1000 mg/kg); phosphate, <200 mg/kg

Essentiality: plants, no; animals, no

General

Barium occurs in relatively high concentrations in the earth's crust. It has only one oxidation state, +2. Barium is physiologically inactive in plants under normal circumstances, but its soluble salts are highly toxic to animals and man. Barium sulfate ($BaSO_4$), which is highly insoluble, is used in the Roentgen technique as a radiation-absorbent contrast material. In soils, inorganic pyrophosphatase activity can be promoted by 30 to 50 μM Ba^{2+}, while iron and copper cations decrease this activity. Soil pH (increasing with decreasing pH) and sulfur content can influence availability.

Plants

With decreasing soil pH, barium availability increases, but with increasing sulfate concentrations due to acid rain, etc., this uptake decreases. For soils high in calcium, magnesium, and sulfur, barium availability decreases. Plants enriched with barium are usually not toxic to grazing animals because its concentration is not sufficiently high. In plant physiological processes, barium is a competitor with calcium and strontium; barium is taken up preferentially against strontium. Barium at 200 mg/kg is considered moderately toxic and in excess of 500 mg/kg toxic. Marine brown algae can significantly concentrate barium.

Animals/Man

Toxic intake is 200 mg, and lethal intake is 3.7 g. The total mass of the element in an average (70-kg) person is 22 mg. Soluble salts of barium are highly toxic and are used to control wild animals (wolf, bear, etc.). Barium is in competition with other alkaline earth metals and can replace calcium in calmodulin-dependent contraction or renal arteries (rabbits), but the needed concentration is much higher. Low concentrations have been found in ancient human bones, suggesting a diet high in meat (plants contain higher barium concentrations). In human tissues, the kidneys may accumulate high concentrations, over 0.1 mg/kg. Daily human intake is estimated as 500 μg/day and urinary excretion between 0.03 to 60 μg/L.

Mobility in Food Chain

Since barium can easily be precipitated as barium sulfate, it is not very mobile and has a very low bioaccumulation index.

BERYLLIUM (Be)

Silvery-white, lustrous, relatively soft metal; unaffected by air and water even in red heat

Atomic Number: 4   **Abundance in Lithosphere**: 2.8 mg/kg
Atomic Weight: 9.01218   **Common Valence State**: Be^{2+}

Common Mineral Forms: beryl (Be-silicate: $Be_3Al_2Si_6O_{18}$), bertrandite $[Be_4S_2O_7(OH)_2]$

Total Content in Soils: 0.1–15 mg/kg; mean value less than 1.6 mg/kg

Soluble Content in Soils: 0.4–1.0 µg/L in soil solution

Content in Sea Water: 5.6 ng/L

Content in Fresh Water: less than 1 ng/L; reference 0.1 µg/L

Chemical Species in Water: Be(OH)

Content in Man: blood, 1.0 ± 0.4 µg/L; bone, 0.0003 mg/kg; muscle, 0.032 mg/kg; urine, 0.4–0.9 µg/L; total mass in average (70-kg) person, 0.036 mg

Content in Plants: 0.001–0.4 mg/kg; reference plant, 0.001 mg/kg

Content in Common Foods: less than 1 mg/kg; lettuce leaves, 0.033 mg/kg; tomato fruit, 0.24 mg/kg

Essentiality: plants, no; animals, no

General

This relatively common element is widely distributed at very low concentrations and binds easily with organic substances. It is enriched in some coals and will accumulate in organic soil horizons.

Plants

Beryllium content in plants is very low, usually less than 1 mg/kg; however, beryllium can accumulate in plant roots, particularly Leguminosae and

Cruciferae, which seem to be beryllium accumulators. Beryllium has been found to be highly toxic to plants at relatively low concentrations in soil solution, with leaf levels of 10 to 50 mg/kg toxic. Beryllium will interfere with the uptake of calcium, magnesium, and phosphorus by plant roots. Beryllium applied to soil as beryllium sulfate at the rate of 200 mg/kg was found to be lethal to collard and wheat plants. Beryllium at levels greater than 2 mg/L in a nutrient solution was found to reduce the growth of alfalfa, lettuce, peas, and soybeans. Beryllium can accumulate in plant roots and partly in shoots, but is rarely found in fruits.

Animals/Man

Daily human intake of beryllium is estimated to be 0.4 µg/day, and urinary excretion is between 0.04 to 0.08 µg/L. Most beryllium compounds are believed to be highly toxic and some forms are carcinogenic. Apparently, beryllium is poorly absorbed through the gut, and thus ingestion is not a hazard. However, when ingested at high concentrations, beryllium can be severely toxic, causing rickets.

Mobility in Food Chain

Beryllium is moderately mobile with only a low bioaccumulation index.

BISMUTH (Bi)

Brittle metal, silvery luster with pink tinge; stable to oxygen and water; dissolves in concentrated nitric acid

Atomic Number: 83   **Abundance in Lithosphere**: 0.048 mg/kg
Atomic Weight: 208.9804   **Common Valence State**: Bi^{3+}

Common Mineral Forms: Bismite (α-Bi_2O_3), bismuthinite (Bi_2S_3), and bismutite [$(BiO)_2CO_3$]

Total Content in Soils: 0.13–0.42 mg/kg

Content in Sea Water: 5.1×10^{-8} mg/L

Content in Fresh Water: reference content 0.05 µg/L

Chemical Species in Water: BiO^+ and $Bi(OH)_2^+$

Content in Man: blood, <3 µg/L; bone, <0.2 µg/g; kidney, 0.4 ± 0.1 µg/g

Content in Animals: 4 µg/kg

Content in Plants: mean <0.02 mg/kg; found at 1–15 mg/kg in some trees; reference plant, 0.01 mg/kg

Content in Common Foods: vegetables, 0.06 mg/kg

Essentiality: plants, no; animals, no

General

The physiological role of bismuth is mostly unknown in spite of the fact that some bismuth-containing compounds are used in medical praxis. Bismuth is enriched in organic sedimentary materials, such as coal and bituminous sediments. Since the bismuth (Bi^{3+}) cation has some similarity with lead (Pb^{2+}), bismuth sometimes occurs in association with lead. The bismuth cation (Bi^{3+}) appears to be poorly adsorbed in soils, although its uptake by plants is generally considered quite easy, but in the presence of some organic compounds and bacteria, this process can be helped markedly.

Plants

Lethal bismuth greater than 27 mg/L for plants.

Animals/Man

Bismuth is a relatively non-toxic element and is found in low concentrations in different human organs and urine (37 µg/L). Dietary intake ranges from 0.005 to 0.002 mg/day.

Mobility in the Food Chain

Bismuth is relatively immobile in the environment and has a low bioaccumulation index. Some bacteria accumulate bismuth, and chelation with some organic compounds increases its mobility in living organisms.

BORON (B)

Non-metal with several allotropes, amorphous boron is a dark powder that is unreactive to oxygen, water, acids, and alkalis; forms metal borides with most metals

Atomic Number: 5   **Abundance in Lithosphere**: 10 mg/kg
Atomic Weight: 110.81   **Common Valence States**: B^{3+} [exists as B_2O_3, H_3BO_3, and $B(OH)_3$]

Common Mineral Forms: borax {$Na_2[B_4O_5(OH)_4] \cdot 8H_2O$], colemanite {$Ca_2[B_3O_4(OH)_3]_2 \cdot 2H_2O$}

Total Content in Soils: 5–80 mg/kg

Soluble Content in Soils: saturated paste, 3060 µg/L; hot-water extractable: deficient when less than 2.0 mg/kg and toxic when greater than 2.5 mg/kg

Content in Sea Water: 4.41 mg/L

Content in Fresh Water: 10 µg/L; >1.1 mg/L toxic to some plants; 0.2–2.2 mg/L toxic to animals

Chemical Species in Water: $B(OH)_3$ or $B(OH)_4^-$

Content in Humans: muscle, 0.33–1 mg/kg; bone, 1.1–3.3 mg/kg; blood, 0.013 mg/dm^3; urine, 0.04–6.6 µg/L; kidney, 0.6 µg/g; milk, 0.27 mg/L

Content in Animals: total body, 0.5 mg/kg; 50–600 µg/kg fresh body weight; milk, cow, 0.33; horse, 0.10; guinea pig, 0.59 mg/L

Content in Plants: range 5–200 mg/kg; 10–200 mg/kg sufficiency range; reference plant, 40 mg/kg

Essentiality: plants, yes; animals, no

General

Boron is the 17th most abundant trace element in the lithosphere and is very mobile, although not evenly distributed. Boron has been intensively studied in terms of its soil and plant chemistry. The role of boron in animal and human nutrition is just now being investigated more intensively.

Plants

Boron is an essential micronutrient. The sufficiency range for many crops is from 10 to 200 mg/kg dry weight of tissue. In general, monocots have a low boron requirement (1 to 6 mg/kg), while dicots have a high requirement (20 to 70 mg/kg), and dicots with a latex system require 80 to 100 mg/kg. Plants sensitive to boron deficiency have a fairly high tolerance for high boron levels. Crops vary widely in their tolerance to boron. It is thought to play a major role in the carbohydrate chemistry of plants. Boron deficiency is likely to occur on sandy low organic matter content soils as well as organic soils.

Animals/Man

Boron is not essential. Dietary intake is 1 to 3 mg, and toxic intake is 4 g. Boron levels greater than 0.2 to 2.2 mg/L in water supplies can be toxic to domestic animals. There is growing evidence that boron may be an essential element, involved in bone formation, and rapidly absorbed and very rapidly excreted through the urine. Calcium may play a role in influencing boron absorption, as is the case with plants.

Mobility in Food Chain

Boron is a highly mobile element and has a very high bioaccumulation index.

CADMIUM (Cd)

Silvery metal, tarnishes in air, soluble in acids but not alkalis

Atomic Number: 48   **Abundance in Lithosphere**: 0.18 mg/kg
Atomic Weight: 112.40   **Common Valence State**: Cd^{2+}

Common Mineral Forms: greenockite (CdS)

Total Content in Soils: 0.01–3.0 mg/kg

Soluble Content in Soils: 0.1–14.0 mg/kg

Content in Sea Water: 1.1×10^{-6} to 38×10^{-6} mg/L

Content in Fresh Water: reference content 0.2 μg/L; 5 mg/kg in water produced systolic hypertension in rats

Chemical Species in Water: Cd^{2+} and $CdOH^+$

Content in Humans: muscle, 0.14–3.2 mg/kg; bone, 1.8 mg/kg; blood, 0.0053 mg/dm^3; urine, 0.02 mg/L; scalp hair, 0.354 mg/kg

Maximum Daily Intake from Toy Materials: 0.6 μg

Content in Animals: 0.1–0.5 mg/kg

Content in Plants: 0.1–1.0 mg/kg; reference plant, 0.05 mg/kg; 3 mg/kg will reduce plant growth

Content in Common Foods: 0.0031 (sweet corn) to 0.078 (peanut) μg/g fresh weight

Essentiality: plants, no; animals, no

General

Cadmium is becoming an element of concern due to its presence in waste products, primarily sewage sludge, that are land disposed. Cadmium loading rates have been established for soils by the U.S. EPA: 20 kg/ha for high cation exchange capacity (CEC) soils with a pH of 6.5 and 5 kg/ha for acid soils. An increase in soil cadmium content will be reflected in a similar increase in crop plant tissues. Cadmium content of surface soils is strongly influenced by man's activity.

Plants

Plants vary in their sensitivity to cadmium in nutrient solutions from 0.2 to 9 mg/kg. Cadmium at 3 mg/kg in plants will depress growth. Cadmium toxicity symptoms are leaf chlorosis and necrosis followed by leaf abscission as cadmium interferes with net photosynthesis and the uptake and transport of mineral elements in the plant. The effect of cadmium on plants varies with species and with other elements, with both synergistic and antagonistic effects. Cadmium content in plants is highly correlated with that found in soils; the degree of cadmium increase varies widely with plant species. Cadmium solubility in soil decreases significantly with an increase in soil pH. When exposed to high levels of cadmium in the rooting environment, some root crops (turnips) and leafy vegetables (spinach) will contain sufficient cadmium to pose a potential health hazard in their consumption. For some seed crops, cadmium is not translocated to the developing seed, although the plant is exposed to high levels of available cadmium in the rooting medium.

Man

Daily dietary intake is 0.007 to 3.0 mg, toxic intake is 30 to 330 mg, lethal intake is 1.5 to 9.0 g, and total mass of the element in an average (70-kg) person is 50 mg. Liquid intake of 16,000 µg/L can cause severe gastrointestinal symptoms. According to the U.S. Bureau of Foods, the mean daily cadmium intake is estimated to be 39 µg. Intake for an adult male is 33 µg and for an adult female 26 µg. The danger level is 200 µg. The safe upper limit(s) for daily cadmium intake set by the World Health Organization (WHO)/Food and Agricultural Organization (FAO) is 57 to 71 µg, and the limit set by the U.S. EPA is 70 µg. The major dietary source of cadmium is cereal grains. Cigarette smoking can account for a daily cadmium intake (normal range 5 to 20 µg/day) of up to 70 µg.

Animals

A level of 45 mg of cadmium in the diet of rats for 6 months will cause slight toxic symptoms, 0.5 mg/day in a total diet is toxic, and 16 mg/day is lethal.

Mobility in Food Chain

The index of bioaccumulation ranges between 1 to 10. Therefore, cadmium will accumulate in some portions of the environment, posing a potential health hazard due to its uptake and accumulation in some food crops.

CESIUM (Cs)

Soft, shiny, gold-colored metal; reacts rapidly with oxygen and explosively with water

Atomic Number: 55   **Abundance in Lithosphere**: 3.2 mg/kg
Atomic Weight: 132.9054   **Common Valence State**: Cs^+

Common Mineral Forms: pollucite [$(Cs, Na)_4Al_4Si_9O_{26} \cdot H_2O$]

Total Content in Soils: 4 mg/kg (range 0.3–26 mg/kg)

Content in Sea Water: 3.0×10^{-4} mg/L

Content in Fresh Water: reference level 0.005 µg/L

Chemical Species in Water: Cs^+

Content in Humans: muscle, 0.07–1.6 mg/kg; bone, 0.013–0.052 mg/kg; blood, 0.0038 mg/dm^3

Content in Animals: 64 µg/kg

Content in Plants: range 0.03–0.4 mg/kg; reference plant, 0.5 mg/kg

Content in Common Foods: 0.2–3.3 µg/g fresh weight in vegetables and <0.1–2.9 µg/g fresh weight in fruits

Essentiality: plants, no; animals, no

General

The geochemisty of cesium is similar to that of rubidium, acting similarly to other monovalent cations. Cesium tends to accumulate in the organic horizons of soils.

Plants

Cesium is not required by plants, although its uptake by roots does not parallel the chemistry of potassium. The addition of lime and organic matter to the soil decreases its uptake by plants.

Animals/Man

Daily dietary intake is 0.004 to 0.03 mg, and cesium is considered a non-toxic element. There is some evidence that cesium may substitute for potassium, with evidence that cesium may be a beneficial element.

Mobility in Food Chain

Cesium has a relatively high bioaccumulation index and would be considered a mobile element within the environment.

CHROMIUM (Cr)

Hard blue-white metal; soluble in HCl and H_2SO_4 but not HNO_3, H_3PO_4, or $HClO_4$ due to formation of protective layer; resists oxidation in air

Atomic Number: 24	**Abundance in Lithosphere**: 200 mg/kg
Atomic Weight: 51.996	**Common Valence States**: Cr^{3+}, Cr^{6+}, CrO_4^{2-}

Common Mineral Forms: chromite ($FeCrO_4$), sometimes this form has high cobalt and nickel contents (chromates); chrochoite ($PbCrO_4$)

Total Content in Soils: 5–1000 mg/kg; mean 65 mg/kg (serpentine soils may contain several percent chromium)

Soluble Content in Soils: 10 µg/L (saturated paste)

Content in Sea Water: 0.16 µg/L; 1–5 mg/L reduces photosynthesis in kelp

Content in Fresh Water: 0.18 µg/L; reference level 1.0 µg/L; 0.03–65 mg hexavalent Cr per liter inhibits algae

Chemical Species in Water: CrO_4^{2-} or $Cr(OH)_3$

Content in Marine Animals: 0.2–1.0 mg/kg

Content in Animals: 75 µg/kg

Content in Humans: muscle, 0.0024–0.84 mg/kg; bone, 0.1–33 mg/kg; blood, 0.006–0.11 mg/dm³; urine, 0.02 mg/L; scalp hair, 0.195 mg/kg

Maximum Daily Intake from Toy Materials: 0.3 μg

Content in Plants: 0.02–0.2 mg/kg; reference plant, 1.5 mg/kg; phytotoxic at >10 mg/kg

Content in Common Foods: range 0.05 mg/kg (apple) to 0.2 mg/kg (wheat grain)

Content in Fertilizers: 2–1000 mg/kg

Content in Air: 0.001–1.0 ng/m³ (in industrial areas 30–50 ng/m³)

Essentiality: plants, no; animals, yes

General

The chemistry of chromium in the soil is fairly complex based on the valent state (+2 to +6) which affects solubility and reactivity. The chromium(III) (Cr^{3+}) cation is the stable form commonly found in the environment, and this form of the element is considered essential for some biological functions. The chromium(VI) (Cr^{6+}) form is toxic to plants, animals, and man, although its occurrence is not common in the environment. Chromium(VI) is 100 times more toxic to plants, animals, and man than trivalent chromium (Cr^{3+}). The solubility of both forms is significantly affected by pH; the lowest solubility occurs between pH 5.5 and 8.0. Plant availability of chromium on high-content soils can be reduced by liming and by the addition of phosphate fertilizers and organic matter.

Plants

Plants vary in their sensitivity to chromium; 5 to 15 mg/kg in nutrient solution results in toxicity, and >150 mg/kg in soil is toxic to some plants. Chromium tends to accumulate in roots and is not easily translocated. There is some evidence of a stimulatory effect of low levels of chromium on plant growth.

Animals/Man

Toxic intake is 200 mg, and lethal intake is >3.0 g. Trivalent chromium (Cr^{3+}) has a lower order of toxicity. Absorption of chromium is influenced by the presence of chelating agents and other metals, particularly zinc and iron, and it is excreted primarily through urine. Chromium is required for normal carbohydrate and lipid metabolism in the body, and its deficiency affects many biological functions. Chromium supplementation improves glucose tolerance and blood lipids. The daily requirement of chromium for adults is 50 to 200 μg.

Mobility in Food Chain

Chromium has a moderate index of bioaccumulation.

COBALT (Co)

Lustrous, silvery-blue, hard metal; ferromagnetic; stable in air, slowly attacked by dilute acids

Atomic Number: 27   **Abundance in Lithosphere**: 20 mg/kg
Atomic Weight: 58.9332   **Common Valence States**: Co^{2+}, Co^{3+}

Common Mineral Forms: smaltite ($CoAs_2$), cobaltine (CoAsS), linnaeite (Co_3S_4)

Total Content in Soils: 1–40 mg/kg

Soluble Content in Soils: 0.10–1.0 mg/kg (0.5 *N* HOAc)

Content in Sea Water: 6.9×10^{-6} mg/L

Content in Fresh Water: 0.01–0.18 mg/L; reference level 0.5 μg/L

Chemical Species in Water: Co^{2+} and $CoCO_3$

Content in Humans: muscle, 0.028–0.65 mg/kg; bone, 0.01–0.04 mg/kg; blood, 0.0002–0.04 mg/dm^3; urine, 0.02 mg/L; scalp hair, 0.195 mg/kg; kidney, 0.5–15 μg/g; milk, 1.3–3.0 mg/L

Content in Animals: 1 mg/kg (average)

Content in Plants: legumes, 0.10–0.57 mg/kg; grass, 0.03–0.27 mg/kg; reference plant, 0.2 mg/kg

Content in Common Foods: 0.04–0.9 mg/kg (range); 8–210 μg/g (range)

Content in Fertilizers: 0.6–12 mg/kg

Essentiality: plants, no; animals, yes

General

The geochemical characteristics of cobalt are similar to iron and manganese, with soil organic matter and clay content determining its distribution and behavior. Alkaline and calcareous soils and soils high in organic matter are associated with cobalt deficiency in grazing animals. In general, the cobalt content of soils is determined primarily by what was in the parent material.

Plants

Cobalt is essential for blue-green algae and microorganisms that fix atmospheric nitrogen (N_2), but it is unclear whether cobalt is essential for plants, although there is evidence of some beneficial effect on some plants. Cobalt concentrations in a nutrient solution ranging from 0.1 to 30 mg/L have been found to be toxic to many plants, with toxic concentrations in plants varying widely from 6 to 143 mg/kg depending on plant species. The symptom of cobalt toxicity is chlorosis. Cobalt can easily be taken up through the roots and then translocated primarily in the transpiration stream.

Animals/Man

Cobalt is a component of vitamin B_{12}, and this is the only known function of cobalt. Toxic intake is 500 mg, and total mass of the element in an average (70-kg) person is 1.5 mg. In 1935, research in Australia established the essentiality of cobalt for ruminants. Cobalt is widely distributed in the body, with high concentrations in liver, bone, and kidney. Grass for ruminants in healthy areas contains 0.1 mg/kg or more, compared with 0.004 to 0.07 mg/kg for deficient areas. Cobalt deficiency has never been clearly demonstrated in monogastric species. The maximum dietary tolerable level of cobalt for common livestock species is 10 ppm.

Mobility in Food Chain

Cobalt has a moderately high bioaccumulation index and is fairly mobile in the food chain.

COPPER (Cu)

Reddish metal, malleable and ductile, with high electrical and thermal conductivities; resistant to air and water but slowly weathers to green patina of carbonate

Atomic Number: 29 **Abundance in Lithosphere**: 70 mg/kg
Atomic Weight: 63.546 **Common Valence State**: Cu^{2+}

Common Mineral Forms: chalcopyrite ($CuFeS_2$), chalcocite (Cu_2S), cuprite (Cu_2O), malachite [$Cu_2(CO_3)(OH)_2$]

Total Content in Soils: 2–100 mg/kg; geometric mean 18 mg/kg

Soluble Content in Soils: <1 mg/kg (3–135 µg/L in soil solution)

Content in Sea Water: 8.0×10^{-5} mg/L

Content in Fresh Water: 0.01–2.8 mg/L; reference level 3.0 µg/L

Chemical Species in Water: $Cu(OH)^+$ or $CuCO_3$

Content in Humans: muscle, 10 mg/kg; bone, 1–26 mg/kg; blood, 1.01 mg/dm^3; kidney, 1.07–4.19 µg/g; urine, 6.1–30.3 µg/L; scalp hair, 26.0 mg/kg

Content in Animals: 2.4 mg/kg; cow's milk, 0.3 mg/kg

Content in Plants: 1–10 mg/kg; reference plant, 10 mg/kg

Content in Common Foods: 3–80 mg/kg; background values (raw product) 0.26–12.0 µg/g wet weight

Essentiality: plants, yes; animals, yes

General

Copper is immobile in soils. It is relatively uniform in profile distribution, although there can be considerable bioaccumulation at the soil surface. Copper can be easily precipitated and interacts readily with both organic and inorganic substances with widely varying solubilities due to pH. Cropland soil copper contamination is not uncommon in areas where copper-containing chemicals were in common use as well as from industrial pollution and/or from land deposition of industrial wastes and sewage sludge.

Decrease in plant availability can be obtained by liming, the addition of phosphate fertilizers, and addition of organic matter to the soil.

Plants

Copper is an essential micronutrient with a sufficiency range of 5 to 30 mg/kg dry weight of tissue. Plant species vary in their tolerance to copper. Toxicity occurs when copper tissue levels exceed 20 to 30 mg/kg. Copper is particularly toxic to roots, with relatively little being translocated to plant tops. The symptom of copper toxicity is chlorosis, as high copper interferes with iron metabolism. Copper deficiency is not common because the requirement for most crops is quite low. Copper deficiencies are most likely to occur on organic soils and on mineral soils with a high pH (>7.5) and/or high (>2%) organic matter content.

Animals/Man

Toxic intake is >250 mg, and total mass of the element in an average (70-kg) person is 73 mg. Normal intake is 2 to 5 mg/day. Water containing 800 µg/L was found to be fatal in a 15-month-old infant. In cattle, high intake of molybdenum can result in copper deficiency, while intake of normal copper and low molybdenum can result in copper toxicity. Sheep are particularly sensitive to copper toxicity; 1.5 g per sheep per day for 30 days will be fatal.

Mobility in Food Chain

Copper has a high bioaccumulation index and can accumulate in plants to high levels, which may pose environmental health problems.

FLUORINE (F)

Pale yellow gas, most reactive of all elements and the strongest commercially available oxidant

---

Atomic Number: 9   **Abundance in Lithosphere**: 950 mg/kg
Atomic Weight: 18.998   **Common Valence State**: F^-

Common Mineral Forms: fluospar (CaF_2), fluoroapatite [$Ca_5(PO_4)_3F$], cryolite (Na_3AlF_6)

Total Content in Soils: <10–1200 mg/kg

Soluble Content in Soils: 10–20 mg/kg

Content in Sea Water: 1.3 mg/L

Content in Fresh Water: 0.1–250 mg/L

Content in Marine Animals: 2 mg/kg, accumulates in sponges and molluscs

Content in Animals: 150–500 mg/kg in soft tissues, 1500 mg/kg in bone

Content in Humans: muscle, 0.05 mg/kg; bone, 2000–12,000 mg/kg; blood, 0.3 mg/dm^3

Content in Plants: 0.2–20 mg/kg; reference plant, 2 mg/kg

Content in Common Foods: <0.1 mg/kg

Content in Fertilizers: 8500–15,500 mg/kg in phosphate fertilizers; phosphorites, 31,000 mg/kg

Essentiality: plants, no; animals, yes

General

Fluorine mobility is considerably complex in soils affected by clay mineral composition, pH, and concentration of calcium and phosphorus, while its soil content is related to that in the parent material. Organic complexes of fluorine are readily available to plants. A major source of fluorine is from air pollution, which can affect plant and animal life through the air and by accumulation in soil.

Plants

High (30 to 300 mg/kg) fluorine levels in plants are primarily due to air pollution. Plant damage is primarily due to fluorine forms as hydrofluoric acid; the extent of damage varies with plant species. Plant uptake from the soil is passive, and fluorine is easily transported in plants, but such uptake is not normally phytotoxic. Normal fluorine background level in plant leaves is generally less than 30 mg/kg.

Animals/Man

Daily dietary intake is 0.3 to 0.5 mg, toxic intake is 20 mg, lethal intake is 2 g, and total mass of the element in an average (70-kg) person is 2.6 g. High (30 to 40 mg/kg) fluorine content in forage plants can be detrimental to animal health, resulting in fluorosis, which is found in many parts of the world. However, fluorine is not equally toxic to all species; ruminants are more susceptible than non-ruminants. The major source of fluorine for humans is water, whereas livestock can receive toxic quantities through contaminated plant materials and phosphate supplements. Ingested fluorine is rapidly absorbed, and the primary pathway of fluorine excretion is through the urine.

Mobility in Food Chain

Fluorine has a slight bioaccumulation index.

GALLIUM (Ga)

Soft, silvery-white metal, stable in air and with water; soluble in acids and alkalis; longest liquid range of all elements

Atomic Number: 31	**Abundance in Lithosphere**: 19 mg/kg
Atomic Weight: 69.72	**Common Valence State**: Ga^{3+}

Common Mineral Forms: germanite (Cu_6FeGaS_4) usually contains 1% Ga; in some bauxites, there is 50 mg/kg Ga, and 1.5% Ga is found in coal-ash

Total Content in Soils: 15 mg/kg (range 2–100 mg/kg)

Content in Sea Water: 3×10^{-5} mg/kg (sea plants, 0.5 mg/kg)

Content in Fresh Water: <1 ng/L; reference level 0.1 µg/L

Chemical Species in Water: $Ga(OH)_4^-$

Content in Humans: muscle, 0.0014 mg/kg; blood, <0.08 mg/dm^3; liver, 0.7 ng/g

Content in Animals: <6 µg/kg

Content in Plants: 0.5 mg/kg

Essentiality: plants, no; animals, no

General

The ionic radii of iron (0.064 nm) and gallium (0.062 nm) are very close; therefore, these two elements have similar, sometimes antagonistic, physiological effects. The transport of both elements is by transferrin. In some cases, gallium promotes the uptake of iron.

Plants

Gallium nitrate over 10 μg/L has been shown to be phytotoxic to blue-green algae. In appropriate concentrations, gallium stimulates the growth of some microorganism (e.g., *Aspergillus niger*). It has been demonstrated that gallium in appropriate concentrations promotes the growth of algae and the biosynthesis of photosynthetic pigments. In the case of *Anacystis nidulans*, oxygen (O_2) production increased and the activity of the key enzyme, fructose-1,6-bisphosphatase, was enhanced. The activity of redox enzymes was also enhanced. In field experiments, gallium nitrate applied to tomato plants resulted in higher yields of better quality, with dry matter content and reducing sugar level enhanced significantly. In this experiment, iron metabolism may have been enhanced as the iron content of the plant was increased. Therefore, gallium may have a beneficial role.

Animals/Man

It has been demonstrated that gallium nitrate added to the diet contributes to bone formation in rats that are on a low-calcium diet. Gallium has been demonstrated as an efficient agent against mammary adenocarcinoma in mice; its cytotoxicity is time dependent. It has been demonstrated that gallium-transferrin chelate inhibits the growth of KL-60 cells because it inhibits the cellular incorporation of iron. Gallium may have a stimulatory effect on iron uptake. Radioactive ^{67}Ga has been used as a marker in many malignant diseases. Gallium has been demonstrated to inhibit cell growth of T-lymphoblastic leukemic cells, as this process is a synergistic interaction with alpha-interferon. Additional research is needed to solve some of the contradictory results in the gallium-iron correlation. Urinary excretion is 0.4 μg/L.

Mobility in Food Chain

Gallium has a low bioaccumulation index. For biological transport, transferrin is the main carrier molecule.

GERMANIUM (Ge)

Ultrapure, a silvery-white brittle metalloid element, stable in air and water, unaffected by acids, except nitric and alkalis

Atomic Number: 32	**Abundance in Lithosphere**: 1.8 mg/kg
Atomic Weight: 72.61	**Common Valence State**: Ge^{4+}

Common Mineral Forms: widely distributed in many minerals (0.3–2.4 mg/kg), recovered as a by-product of zinc and copper refining

Total Content in Soils: 0.6–2.1 mg/kg (mean 1.1 mg/kg)

Content in Sea Water: $0.007–0.14 \times 10^{-6}$ mg/L; phytoplankton accumulate Ge

Content in Fresh Water: 0.1 µg/L

Chemical Species in Water: $Ge(OH)_4$

Content in Humans: muscle, 0.14 mg/kg; blood, 0.2 µg/g; liver, 0.04 µg/g; kidney, 9 ng/g

Content in Animals: 0.01–0.1 mg/kg

Content in Plants: <0.1 mg/kg; reference plant, 0.01 mg/kg

Content in Common Foods: grains, 0.09–0.7 mg/kg (fresh weight); vegetables, 0.02–1.07 mg/kg (fresh weight)

Essentiality: plants, no; animals, no

General

Silicon and germanium have similar chemistries due to the fact that both are in Group IV-A of the periodic table, have similar silicon-hydrogen and germanium-hydrogen bonding energies (63 and 56.7 kcal/mol, respectively), and similar atomic radii (0.134 nm for silicon and 0.139 nm for germanium).

Plants

Germanium is toxic to plants at concentrations in solution greater than 40 μM; primary leaves become necrotic with elevated peroxidase activity. Silicon accumulator plants will accumulate germanium as compared to low silicon accumulator plants. It has been observed that 5 mg/kg GeO_2 is toxic to rice plants. Blue-green algae and diatoms will accumulate germanium, as will some bacteria (*Pseudomonas* spp., *Oscillitoria* spp., and *Spirulina* spp.).

Animals/Man

Daily dietary intake is 0.4 to 1.5 mg. Germanium is non-toxic and is easily and rapidly excreted through urine. Some germanium compounds are used as pharmaceutical agents.

Mobility in Food Chain

Germanium has a moderate bioaccumulation index.

GOLD (Au)

Soft metal with characteristic yellow color; highest malleability and ductibility of any element; unaffected by air, water, acids (except HNO_3-HCl), and alkalis

Atomic Number: 79   **Abundance in Lithosphere**: 0.0011 mg/kg
Atomic Weight: 196.967   **Common Valence State**: Au^+

Common Mineral Forms: occurs as metal

Total Content in Soils: 1–2 ng/kg

Content in Sea Water: 1×10^{-5} mg/L

Content in Fresh Water: reference level 0.002 mg/L

Chemical Species in Water: $Au(OH)_4^-$

Content in Marine Animals: 0.3–5.0 µg/kg

Content in Humans: bone, 0.016 mg/kg; blood, $(0.1–4.2) \times 10^{-5}$ mg/dm^3

Content in Animals: 0.5 µg/kg

Content in Plants: 0.0005–0125 mg/kg; reference plant, 0.001 mg/kg

Content in Common Foods: 0.01–0.4 µg/kg

Essentiality: plants, no; animals, no

General

Gold is a rare element and is transported in soils mainly as organometallic compounds or chelates.

Plants

Gold, when taken up by plants, is easily transported to plant tops. Some plants are high accumulators of gold. It can be toxic at high concentrations, resulting in necrosis and wilting.

Animals/Man

Gold is non-toxic. Gold compounds have been tested for pharmaceutical use for treating various diseases.

Mobility in Food Chain

It would be expected that gold would not easily move or accumulate in the environment.

HAFNIUM (Hf)

Lustrous, silvery, ductile metal; resists corrosion due to oxide film, but powdered Hf will burn in air; unaffected by acids (except HF) and alkalis

Atomic Number: 32	**Abundance in Lithosphere**: 1.8 mg/kg
Atomic Weight: 72.61	**Common Valence State**: Hf^{4+}

Common Mineral Forms: has no known minerals of its own; zirconium minerals and baddeyite contain 2% Hf

Total Content in Soils: 0.5–20.0 mg/kg (mean 6 mg/kg)

Content in Sea Water: 8 ng/L

Content in Humans: blood, 6 ng/L; tooth enamel, 80 ng/kg

Content in Animals: less than 20 µg/kg

Content in Plants: less than 20 µg/kg; reference plant, 0.05 mg/kg

Essentiality: plants, no; animals, no

General

Hafnium was discovered in 1922 by Hungarian scientist G. Hevesy (Nobel Prize winner) and D. Coster, a Dutch scientist. Hafnium is chemically very similar to zirconium, but its mobility is less than that of zirconium. Hafnium easily forms organic chelates with humic compounds in the soil.

Plants

Hafnium content is probably less than 20 µg/kg; sea plants may accumulate hafnium to 10 µg/kg.

Animals/Man

The zirconium/hafnium ratio in animal and human diets is usually 35, similar to the level in the geosphere. The average concentration in human blood is estimated as 6 µg/L and in tooth enamel as 80 ng/g. The average intake of hafnium by humans is 1 µg, and urinary excretion is estimated to be 0.01 to 0.8 µg/L. Hafnium is generally considered to be non-toxic.

Mobility in Food Chain

Some compounds of hafnium have solubilities and mobility similar to zirconium compounds.

INDIUM (In)

Soft, silvery-white metal; stable in air and water; dissolves in acids

Atomic Number: 49	**Abundance in Lithosphere**: 0.049 mg/kg
Atomic Weight: 114.82	**Common Valence State**: In^{3+}

Common Mineral Forms: no known In-containing minerals exist; In is found in some zinc-bearing minerals (sphalerites); some coal-ash materials contain high amounts (50 mg/kg) of In

Total Content in Soils: <0.02–0.5 mg/kg (average 0.2 mg/kg)

Content in Sea Water: 1×10^{-7} mg/L

Content in Man: muscle, 0.0015 mg/kg; blood, 3 µg/L; kidney, 0.03 ± 0.05 µg/g

Content in Plants: 30–710 µg/kg (mean 210 µg/kg); reference plant, 0.001 mg/kg

Essentiality: plants, no; animals, no

General

Indium is primarily found in coals and crude oil and has been reported to be combined with organic substances, associated with soil organic matter, and therefore concentrated in the soil surface. Some sludges may be a source of indium. Near zinc and lead smelters, over 10 mg/kg has been found in the soils of Japan.

Plants

Uptake of indium by plants is pH dependent, and indium toxicity can occur, with symptoms similar to those for aluminum. Indium has been found to stimulate cell growth in cell cultures.

Animals/Man

Indium is relatively non-toxic when compared to other trace elements such as scandium, gallium, chromium, and yttrium, It exhibited a lower incidence of tumors when applied in drinking water. Indium can be easily absorbed from feeds, but a part of the indium absorbed is excreted through the urine. Indium given orally showed a lower toxicity at 30 mg/day, but a 10-mg injection of indium has proven lethal to mice. The metabolism of indium shows some similarity to gallium and iron (Fe^{III}). Indium compounds have been found to be effective against sleeping sickness. Lethal intake is 30 mg.

Mobility in Food Chain

Not known.

IODINE (I)

Gray-black or violet-black, shiny non-metal solid; sublimes easily

Atomic Number: 53   **Abundance in Lithosphere**: 0.14 mg/kg
Atomic Weight: 126.90457   **Common Valence State**: I^-

Common Mineral Forms: brines; Chilean nitrates contain up to 0.3% calcium iodate; seaweed

Total Content in Soils: 0.1–40 mg/kg; mean 2.8 mg/kg

Content in Sea Water: 0.049 mg/L

Content in Fresh Water: reference level 3.0 µg/L

Chemical Species in Water: I^- and IO_3^-

Content in Marine Plants: brown algae, 55–8800 mg/kg

Content in Humans: muscle, 0.05–0.5 mg/kg; bone, 0.27 mg/kg; blood, 0.057 mg/dm^3

Content in Animals: 0.43 mg/kg

Content in Plants: range 2–6 mg/kg; reference plant, 3 mg/kg

Content in Common Foods: 0.01–1.6 mg/kg

Essentiality: plants, no; animals, yes

General

Iodine has been recognized as an essential element for animals and man for two centuries. A number of areas in the world are deficient in this element, resulting in significant health effects. Iodine compounds are soluble and leached from many soils, and retention is associated with organic matter, carbon, and clay contents. High iodine levels exist in alkali soils of arid and semi-arid regions. Release from sea water can be a significant source of iodine, as well as coal burning and deposition of sewage sludge.

Plants

Iodine concentrations greater than 0.5 mg/L in nutrient solution have been found to be toxic to plants, while a low iodine level (0.1 mg/L) has shown some stimulatory effect. There seems to be a lack of a strong association between soil and plant iodine levels. Iodine can be easily absorbed through the leaves. Vegetables contain higher levels of iodine than do other plant types.

Animals/Man

Toxic intake is 2 mg, lethal intake is 35 to 350 mg, and total mass of the element in an average (70-kg) person is 12 to 20 mg. In man, lack of iodine affects the thyroid gland and will result in the formation of goiter. Safe and required daily intake by humans is 0.1.5 mg/day. Iodine is easily absorbed in the human body and excreted through the urine. The requirement for iodine among various animals is fairly consistent (0.10 to 0.50 mg/kg), while sensitivity to high iodine levels varies greatly (5 to 400 mg/kg).

Mobility in Food Chain

Iodine has a slight bioaccumulation index.

IRON (Fe)

When pure, is lustrous, silvery, and soft; rusts in damp air; dissolves in dilute acids

Atomic Number: 26	**Abundance in Lithosphere**: 41,000 mg/kg
Atomic Weight: 55.847	**Common Valence States**: Fe^{2+}, Fe^{3+}

Common Mineral Forms: hematite (Fe_2O_3), magnetite (Fe_3O_4), siderite ($FeCO_3$)

Total Content in Soils: 38 g/kg

Soluble Content in Soils: saturated paste, 50 µg/L; range at common pH levels, 30–550 µg/L

Content in Sea Water: 1×10^{-4} to 4×10^{-4} mg/L

Content in Fresh Water: 0.04–6200 mg/L; reference level 500 µg/L

Chemical Species in Water: $Fe(OH)_2^+$ in oxygen containing, Fe^{2+} under reducing conditions

Content in Marine Animals: 400 mg/kg

Content in Humans: muscle, 180 mg/kg; bone, 3–380 mg/kg; blood, 447 mg/dm^3; milk, 0.26 mg/L; urine, 0.081 mg/L; scalp hair, 10.1 mg/kg

Content in Animals: 40–160 mg/kg

Content in Plants: 20–100 mg/kg (soluble 10 mg/kg); reference plant, 150 mg/kg

Content in Common Foods: 8–40 mg/kg

Essentiality: plants, yes; animals, yes

General

Iron is a major constituent of the lithosphere. Its geochemistry is very complex, depending on its valence state. Iron is associated with various physicochemical conditions. In soil, iron exists mainly as oxides and hydroxides and will be chelated by organic matter. Iron in soil solution is affected by soil pH, decreasing with increasing pH. Under anaerobic conditions, ferric (Fe^{3+}) is reduced to ferrous (Fe^{2+}), which significantly increases its solubility in soils and under acid soil conditions induces possible iron toxicity. For humans and plants, the essentiality of iron was demonstrated over two centuries ago.

Plants

Iron is an essential micronutrient. Its sufficiency range in most plants is 50 to 500 mg/kg dry weight; the critical level is 50 mg/kg for a wide range of plants. Total iron, however, is not usually a reliable indicator of sufficiency, and extractable iron is frequently used for iron status assessment, where 20 to 25 mg/kg is the critical range, depending on the extraction procedure used and plant species. An accurate assessment of the iron status of a plant can be significantly affected by the presence of extraneous iron on tissue surfaces due to aerial deposition and soil contamination as well as contact with iron-containing devices used for tissue preparation. In many instances, indirect measurement, such as chlorophyll content determination, may be a more consistent evaluator of the iron status of the plant. Iron deficiency affects many crops and may be the most frequently occurring deficiency worldwide. Deficiency is mainly associated with alkaline soil conditions, referred to as *lime chlorosis*. Iron deficiency may also be genetically controlled, as some plants have been classed as either iron efficient or iron inefficient. So-called iron-efficient plants are able to acidify the rhizosphere and release iron-chelating substances called siderophores. Iron deficiency can be induced by high zinc and the presence of the bicarbonate (HCO_3^-) anion in the soil solution. Once it occurs, iron deficiency in a plant is difficult to correct even with frequent foliar applications of iron-containing compounds, whether chelated or not. The biochemistry of iron is very complex, and its soil and plant chemistry has been studied intensively.

Animals/Man

Daily dietary intake is 6 to 40 mg, toxic intake is 200 mg, lethal intake is 7 to 35 g, and total mass of the element in an average (70-kg) person is 4.2 g. From 60 to 70% of the iron in the animal and human body is found in the blood; hemoglobin contains 0.35% iron. The most important iron carrier is transferrin, which also plays a major role in metallic ion movement in the body. Ferritin is the main factor in iron storage, while that found in bone marrow affects iron metabolism. Iron deficiency anemia is one of the ten most common human deficiencies, and the effects of iron supplementation are significantly influenced by many biochemical factors. The uptake and mobility of iron depend on a number of different parameters, including effects due to other metals (such as copper, zinc, and titanium) and the presence of chelating compounds.

Mobility in Food Chain

Iron has a low bioaccumulation index.

LEAD (Pb)

Soft, weak, ductile, dull grey metal; tarnishes in moist air but stable to oxygen and water; dissolves in nitric acid

Atomic Number: 82	**Abundance in Lithosphere**: 14 mg/kg
Atomic Weight: 207.2	**Common Valence State**: Pb^{2+}

Common Mineral Forms: galena (PbS), anglesite ($PbSO_4$), pyromorphite [$Pb_5(PO_4)_3Cl$], mimetesite [$Pb_5(AsO_4)_3Cl$]

Total Content in Soils: 3–189 mg/kg (natural background level 10–67 mg/kg, mean 32 mg/kg); considered phytotoxic in soils at 100–400 mg/kg

Soluble Content in Soils: saturated paste, 5.0 µg/L

Content in Sea Water: 30×10^{-6} mg/L (surface); 4.0×10^{-6} mg/L (deep)

Content in Fresh Water: 0.01–5.6 mg/L; reference level 3.0 µg/L

Content in Humans: muscle, 0.23–3.3 mg/kg; bone, 3.6–30 mg/kg; blood, 0.21 mg/dm^3

Maximum Daily Intake from Toy Materials: 0.7 µg

Content in Animals: 2 mg/kg

Content in Plants: reference plant, 1.0 mg/kg

Content in Fertilizers: 2–225 mg/kg

Content in Common Foods: raw agricultural crops, 0.002–0.10 µg/g wet weight (unpolluted areas)

Essentiality: plants, no; animals, no

General

Lead is one of the well-known toxic heavy metals and is a major pollutant. It is primarily introduced into the atmosphere by the use of lead-containing gasoline and then introduced into the food chain by deposition on crop plants and soil dust inhalation. Lead is the least mobile of the heavy metals in soils. It accumulates primarily on the surface, where its increasing presence may begin to affect soil microflora. Lead availability in soil is influenced by pH, decreasing with increasing pH of the soil. Lead is not readily soluble in water and is found in relatively low concentrations. The maximum permissible level in drinking water is 0.05 mg/L.

Plants

Lead at 30 mg/L in nutrient solution has been found to be toxic to plants, with 10 mg/L slowing plant growth and 100 mg/L being lethal. In some types of plants, lead can be as high as 350 mg/kg in plant tissue without visible harm. Extremely low levels (2 to 6 µg/kg) of lead may be necessary for plants, as there is some evidence of a stimulatory effect at low concentrations. Lead can be readily absorbed by plant roots (the amount absorbed greatly varies with plant type), but little (less than 3%) is translocated to the tops. Background lead level in grasses is 2.1 mg/kg and 2.5 mg/kg in clovers. Leafy vegetables, such as lettuce, have a high bioaccumulation character for lead.

Animals/Man

Daily dietary intake is 0.06 to 0.5 mg, toxic intake is 1 mg, lethal intake is 10 g, and total mass of the element in an average (70-kg) person is 120 mg (stored in the skeleton). Lead is both carcinogenic and teratogenic. It adversely affects young (2 to 3 years old) children, primarily impacting their intellectual development. Lead primarily enters the body by soil dust inhalation, although it can enter through the skin or be ingested through contaminated food.

Mobility in Food Chain

Lead has a moderate bioaccumulation index, and its movement within the food chain is due primarily to man's activity.

LITHIUM (Li)

Soft, white, silvery metal; reacts slowly with oxygen and water

Atomic Number: 3 **Abundance in Lithosphere**: 20 mg/kg
Atomic Weight: 6.941 **Common Valence State**: Li^+

Common Mineral Forms: lepidolite ($KLiAl_2Si_3O_{10}$), litiofillate ($LiMnPO_4$), and in different micas

Total Content in Soils: 1.2–98 mg/kg; mean value 30 mg/kg

Soluble Content in Soils: 111 µg/L (saturated paste)

Content in Sea Water: 0.18 mg/L

Content in Fresh Water: 1.1 µg/L (in various geographical regions, mineral waters may be relatively rich in Li); reference level 3 µg/L

Chemical Species in Water: Li^+

Content in Marine Animals: 1 mg/kg

Content in Humans: muscle, 0.023 mg/kg; blood, 0.004 mg/dm^3

Content in Animals: 0.01 mg/kg

Content in Plants: 0.2–6.0 mg/kg; reference plant, 0.2 mg/kg

Content in Common Foods: <4–28 mg/kg

Essentiality: plants, no; animals, possibly yes

General

Lithium is an alkaline metal which shows similarities to sodium and potassium; therefore, antagonistic symptoms can occur. There are no data that would support lithium essentiality for plants, but it does have considerable physiological effects for animals and humans, particularly humans, and therefore may be considered as possibly essential.

Plants

The lithium content of plants varies considerably depending on the soil level, with alfalfa, rye, and red clover considered good indicators of lithium supplementation. Lettuce is a high accumulator of lithium, while apple, onion, and grain crops are generally low in lithium. Lithium in the soil in excess of 2 to 5 mg/kg was toxic to citrus.

Animals

Experiments with goats have suggested that lithium is an essential element for animals, as goats on a lithium-deficient diet had lower body weight gain, lower conception rates, higher abortion rates, and significantly higher mortality rates. Lithium has been found to be toxic to piglets receiving 30 to 60 mg/kg in their diet; poultry (broilers) also is similarly affected by high lithium intake.

Man

In humans, lithium has been found to have successful therapeutic effects in psychiatry for several manias. Average dietary intake is 0.1 mg/day and urinary excretion is 0.5 µg/L. Daily dietary intake is 0.1 to 2 mg, toxic intake is 92 to 200 mg, and total mass of the element in an average (70-kg) person is 0.67 mg.

Mobility in Food Chain

Lithium has a slight bioaccumulation index.

MANGANESE (Mn)

Hard, brittle, silvery metal; reactive when impure and will burn in oxygen; surface oxidation occurs in air; will react with water, dissolves in dilute acids

---

Atomic Number: 25
Atomic Weight: 54.9380

Abundance in Lithosphere: 1000 mg/kg
Common Valence States: Mn^{2+}, Mn^{3+}, Mn^{4+}

Common Mineral Forms: chief ores are pyrolusite [MnO_2], psilomelane or wad [impure MnO_2], cryptomelane [KMn_8O_{16}], manganite [$MnO(OH)$]

Total Content in Soils: 200–3000 mg/kg; mean 545 mg/kg

Soluble Content in Soils: saturated paste, 170 µg/L; range 25–2200 µg/L

Content in Sea Water: $0.4–1.0 \times 10^{-4}$ mg/L

Content in Fresh Water: 0.1–110 mg/L; reference level 5 µg/L

Chemical Species in Water: Mn^{2+}

Content in Marine Animals: 1–60 mg/kg (highest in sponges, lowest in fish)

Content in Animals: 2–4 µg/g; 0.01–0.02 mg/L in milk

Content in Humans: muscle, 0.2–2.3 mg/kg; bone, 0.2–100 mg/kg; blood, 0.0016–0.075 mg/dm^3; milk, 0.01 mg/L

Content in Plants: 10–500 mg/kg, although levels may be as high as 1500 mg/kg without harm to some plant; sufficiency range for most plants, 20–300 mg/kg; reference plant, 200 mg/kg

Content in Common Foods: 2–80 mg/kg; 1–38 mg/kg wet weight in raw agricultural products

Essentiality: plants, yes; animals, yes

General

Manganese is one of the most abundant trace elements in the lithosphere. It exists in the soil in a wide range of oxidation states, depending on physiochemical conditions, and its solubility is dependent on soil pH and redox potential. Manganese availability increases significantly with a decrease in soil pH. Availability is also affected by the presence of organic ligands for neutral to alkaline soils. Manganese is not evenly distributed in the soil profile.

Plants

Manganese is an essential element for plants. The sufficiency range in plant tissue varies from 10 to 500 mg/kg dry weight depending on plant species.

Plants vary considerably in their sensitivity to manganese, and plants sensitive to deficiency are equally sensitive to toxicity. Some plants are high accumulators of manganese without detrimental effects. In general, as the soil pH increases, the availability of manganese decreases, as manganese deficiency is related to high soil pH (>7.5) and toxicity is associated with low soil pH (<5.5). Manganese availability is also reduced by low soil temperature and increasing soil organic matter content. Manganese deficiency symptoms occur primarily on emerging leaves and can be frequently confused with iron deficiency. Manganese toxicity can occur on acid soils (pH <5.5) and under anaerobic conditions. It accumulates as specks of manganese oxide (MnO_2) in the epidermis, giving a symptom referred to as *measles*. The manganese status of a plant can be significantly influenced by its relationship to other elements, primarily iron, as well as phosphorus, calcium, magnesium, and silicon.

Animals/Man

Daily dietary intake is 0.4 to 10 mg, toxic intake is 10 to 20 mg (rats), and total mass of the element in an average (70-kg) person is 12 mg. Manganese is widely distributed throughout the body but in low concentrations. Effects on reproduction are the first signs of manganese deficiency, which also affects skeletal development. Manganese is normally considered the least toxic of the trace elements for poultry and mammals, with maximum tolerable dietary levels varying from 400 to 2000 ppm depending on the species. Absorbed manganese is primarily excreted through feces; little is excreted in urine.

Mobility in Food Chain

Manganese has a low bioaccumulation index.

MERCURY (Hg)

Liquid silvery metal; stable with air and water, unreactive to acids (except concentrated HNO_3) and alkalis

---

Atomic Number: 80   **Abundance in Lithosphere**: 0.05 mg/kg
Atomic Weight: 200.59   **Common Valence State**: Hg^{2+}

Common Mineral Forms: cinnabar (HgS)

Total Content in Soils: 0.01–1.0 mg/kg; mean 0.3 mg/kg

Soluble Content in Soils: saturated paste, 2.4 µg/L

Content in Sea Water: 4.9×10^{-7} mg/L

Content in Fresh Water: <0.1–6.0 µg/kg; reference level 0.1 µg/L

Chemical Species in Water: $Hg(OH)_2$ and $HgOHCl$

Content in Humans: blood, 3–11 µg/L; bone, 0.04–1.04 mg/kg; liver, 0.0019–0.14 mg/kg

Maximum Daily Intake from Toy Materials: 0.5 µg

Content in Animals: 40–50 µg/kg

Content in Plants: 0.5–100 µg/kg; reference plant, 0.1 µg//L

Content in Fertilizers: 0.01–1000 mg/kg

Content in Common Foods: 2.6–86 µg/kg; cereal grains, 0.2–82 µg/kg (mean range 0.9–21 µg/kg); meat, fish, and poultry, 0.017 µg/kg fresh weight

Essentiality: plants, no; animals, no

General

The upper limit for mercury in drinking water set by the WHO is 1 µg/L and that set by the U.S. EPA is 2 µg/L. Much of the mercury found in the environment has been the result of man's activity with contaminated soils ranging from 0.1 to 40 mg Hg/kg. The most toxic forms of mercury are volatile and methylated.

Plants

Mercury can be easily taken up by plants and translocated, binding itself readily with amino acid sulfur atoms. Mercury can be highly toxic to plants. Symptoms are stunting of seedling growth and root development and inhibition of photosynthesis. The mercury content of the edible portions of plants grown on mercury-contaminated sites will range from 0.05 to 37 mg/kg. Mushrooms and lichens are highest, while grains and fruits are lowest.

Animals/Man

The joint WHO/FAO tolerable weekly intake of mercury for a 60-kg man is 43 µg; for methyl mercury, the daily upper limit is 29 µg. Daily dietary intake is 0.004 to 0.002 mg, toxic intake is 0.4 mg, and lethal intake is 150 to 300 mg. Methylated mercury is most toxic.

Mobility in Food Chain

Mercury enters the food chain primarily through atmospheric deposition from coal combustion, smelting, and volcanic activity and from the use of some types of pesticides. Volatilized and methylated mercury compounds are the most toxic. Estimated dietary intake is 2.89 µg/day, with practically all the mercury entering the diet through the consumption of meat, fish, and poultry.

MOLYBDENUM (Mo)

Metal is lustrous, silvery, and fairly soft when pure

Atomic Number: 42   **Abundance in Lithosphere**: 1.5 mg/kg
Atomic Weight: 95.94   **Common Valence State**: Mo^{6+}, Mo^{2+}

Common Mineral Forms: molybdenite (MoS_2), wulfenite ($PbMoO_4$), molybdophyllite ($PbMoSiO_4$)

Total Content in Soils: 0.5–40 mg/kg (mean 2 mg/kg)

Soluble Content in Soils: saturated paste, 730 µg/L

Content in Sea Water: 0.01 mg/L

Content in Fresh Water: 0.3 µg/L; reference level 1.0 µg/L

Chemical Species in Water: MoO_4^{2-}

Content in Humans: muscle, 0.018 mg/kg; bone, <0.7; blood, ~0.001 mg/dm³; milk, 0.02 mg/L

Content in Animals: 0.1–0.2 mg/kg; milk, 0.02 mg/L

Content in Plants: 0.1–3.0 mg/kg; reference plant, 0.5 mg/kg

Content in Phosphorus Fertilizers: 14 mg/kg; rock phosphate, 2–21 mg/kg

Content in Common Foods: 0.013–2.1 µg/g raw agricultural product

Essentiality: plants, yes; animals, yes

General

Molybdenum mainly exists in the anionic form (as molybdates), but sometimes can be found in cationic form; the oxidation number is either Mo^{6+} or Mo^{2+}. Molybdenum is the only essential element in the second line among the transition metals, as it is essential for both plants and animals.

Plants

Molybdenum is an essential micronutrient, with a sufficiency range of 0.1 to 2.0 mg/kg dry weight of tissue. It is involved in nitrogen metabolism of plants [essential for the conversion of the ammonium (NH_4^+) cation to the nitrate (NO_3^-) anion]. When molybdenum is deficient, plants frequently have the appearance of being nitrogen deficient. The molybdenum content of a plant varies considerably with plant species, ranging from 0.07 to 2.5 mg/kg; leguminous plants are higher in molybdenum than grasses. Availability, and therefore content, in plants increases as the soil pH increases from low acidity (<pH 6.0) to neutrality (>pH 7.0).

Animals

Molybdenum requirements are very low, as it is easily absorbed and excreted; however, its activity is closely related to that of copper, and a well-established relationship exists between the two elements. High molybdenum intake can result in a copper deficiency known as *molybdenosis*, while low molybdenum intake can lead to copper toxicity. The effect of this molybdenum-copper relationship varies with animal species, breed, and sex. There is also a molybdenum-sulfur relationship, although it is not as significant as that between molybdenum and copper.

Man

Daily dietary intake is 0.05 to 0.35 mg, toxic intake is 5 mg, and lethal intake is 50 mg (rats).

Mobility in Food Chain

Molybdenum has a moderately high bioaccumulation index.

NICKEL (Ni)

Silvery-white metal, lustrous, malleable, and ductile; resists corrosion; soluble in acids, except concentrated HNO_3, unaffected by alkalis

Atomic Number: 28	**Abundance in Lithosphere**: 75 mg/kg
Atomic Weight: 58.71	**Common Valence State**: Ni^{2+} (rarely Ni^{3+})

Common Mineral Forms: nickeline (NiAs), millerite (NiS), monenozite ($NiSO_4 \cdot 7H_2O$), genthite [$Ni_4(Mg)Si_3O_{10}$]; in olivine, Ni content is 3 g/kg of mineral

Total Content in Soils: 1–200 mg/kg; mean value 20 mg/kg

Soluble Content in Soils: 2 mg/kg (0.5 *N* HOAc); saturated paste, 20 µg/ L; range 3–25 µg/L

Content in Sea Water: 236 ng/L

Content in Marine Plants: 3 mg/kg, plankton are Ni accumulators

Content in Fresh Water: 10 µg/L; reference level 0.3 µg/L

Chemical Species in Water: Ni^{2+}, also $NiCO_3$

Content in Humans: muscle, 1–2 mg/kg; bone, <0.7 mg/kg; blood, 0.01– 0.05 mg/dm^3

Content in Animals: 0.8 mg/kg

Content in Plants: pasture plants, 0.3–3.5 mg/kg; reference plant, 1.5 mg/ kg

Content in Common Foods: raw agricultural product, 0.07–4.8 µg/g wet weight

Essentiality: plants, has been suggested as required; animals, yes

General

For many decades, nickel was regarded as a potentially toxic element, since its concentration in various foods was higher than that needed for living organisms. More recently, it is now considered a possible essential element for plants, although deficiencies can occur under certain circumstances. However, nickel can be toxic at high concentrations and can be a problem in some soils. Liming is one means of reducing nickel availability. Soils high in nickel content are usually due to man's activity.

Plants

Recent research reports suggest that nickel is an essential micronutrient, with 100 µg/kg as the critical concentration in barley plant tissue. Nickel in barley grain at less than 100 ng/kg significantly reduces seed germination and less than 50 ng/kg reduces germination by 70%. The nickel content in the soil is highly correlated with that in the plant, as nickel is readily and rapidly taken up by plants and is highly mobile in plants. There is increasing concern about nickel toxicity because nickel is readily absorbed by plants through the roots as well as from airborne deposition.

Animals

Nickel was first suggested as essential in the early 1970s. Recently established dietary requirements are 50 µg/kg for rats and chicks and >100 µg/kg for ruminants. Most feeds contain sufficient nickel to meet the dietary requirements of ruminants; therefore, deficiency is not likely to occur under normal conditions.

Man

The average human intake of nickel is estimated to be 0.3 to 0.5 mg/day and urinary excretion is 17 µg/day. Toxic intake is 50 mg (rats). Total mass of the element in an average (70-kg) person is 1 mg. Nickel compounds or metal jewelry can cause nickel dermatitis; 8 to 10% of the population experiences this allergic reaction.

Mobility in Food Chain

Nickel has a moderate bioaccumulation index.

NIOBIUM (Nb)

Shiny silvery metal, soft when pure; resists corrosion due to oxide fiim; attacked by hot, concentrated acids but resists fused alkalis

Atomic Number: 41	**Abundance in Lithosphere**: 20 mg/kg
Atomic Weight: 92.906	**Common Valence States**: Nb^{4+}, Nb^{5+}

Common Mineral Forms: niobite (columbite) $(Fe,Mn)Nb_2O_3$, ilmenorutile $(Fe,Mn)Nb_2O_3 + TiO_2$; in some bauxites, 30–70 mg Nb/kg

Total Content in Soils: 5–100 mg/kg, mean value 14 mg/kg

Content in Sea Water: 9×10^{-9} mg/L

Content in Fresh Water: 5 mg/L

Content in Humans: blood, 0.005 mg/dm^3; bone, <0.07 mg/kg; liver, 0.1 µg/g

Content in Animals: 6–8 µg/g

Content in Plants: 3–3000 µg/kg; reference plant, 0.05 mg/kg

Content in Common Foods: vegetables, 10 µg/kg

Essentiality: plants, no; animals, no

General

Niobium closely resembles tantalum and is associated with iron, titanium, and zirconium in its geochemistry; it is a very stable element in the earth's crust. It is slightly soluble in both acid and alkaline conditions and would be moderately available to plants.

Plants

Niobium is relatively common in plant tissues, vegetables, and fruits (e.g., 320 mg/g was found in banana fruit). Plant uptake from the soil is relatively easy in the presence of chelating agents, such as citric, tartaric, or oxalic acids.

Animals/Man

Daily dietary intake is 0.02 to 0.6 mg. It is moderately toxic. Little is known about the metabolism of niobium in animals and man.

Mobility in Food Chain

Unknown.

PLATINUM (Pt)

Silvery-white metal, lustrous, malleable, ductile; unaffected by oxygen and water, only dissolves in *aqua regia* and fused alkalis

Atomic Number: 78	**Abundance in Lithosphere**: ~0.001 mg/kg
Atomic Weight: 195.08	**Common Valence State**: Pt^{2+}

Common Mineral Forms: platinum ores; extracted as a by-product of Cu and Ni refining

Total Content in Soils: 20–75 µg/kg

Content in Sea Water: 1.1×10^{-7} mg/L

Content in Humans: blood, 0.4–1.8 µg/L; urine, <3 µg//L

Content in Plants: 12–56 µg/kg; reference plant, 0.05 µg/kg

Essentiality: plants, no; animals, no

General

Since platinum is a chemically inert element, it is unlikely that it would be of significant environmental concern or a health hazard. Its use in automobile catalytic converters poses no known health hazard at this time.

Plants

In hydroponic studies, platinum was found to accumulate in the roots of plants, with little translocated to the upper portions of the plant.

Animals/Man

Not exactly known.

Mobility in Food Chain

Unknown.

RUBIDIUM (Rb)

Very soft metal with silvery white luster when cut; ignites in air and reacts violently with water

Atomic Number: 37	**Abundance in Lithosphere**: 90 mg/kg
Atomic Weight: 85.4678	**Common Valence State**: Rb^+

Common Mineral Forms: granite and gneiss (Rb-oncosite)

Total Content in Soils: <20–210 (mean range 50–120 mg/kg)

Content in Sea Water: 0.12 mg/L

Content in Fresh Water: 1.0 µg/L (in middle Europe drinking water, 8.1 µg/L); reference level 1 µg/L

Content in Marine Animals: 20 mg/kg

Content in Humans: blood, 1.97 mg/L; bone, 0.1–5 mg/kg; liver, 2.9–6.3 µg/g; total mass in the body, 680 mg

Content in Animals: 1–25 mg/kg

Content in Plants: 0.5–70 mg/kg; reference plant, 50 mg/kg; can be very high (100 mg/kg) in fungi

Content in Fertilizers: phosphate fertilizers, 5 mg/kg

Content in Common Foods: corn and cereal grains, 3–4 mg/kg; onion, 1 mg/kg

Essentiality: plants, no; animals, no

General

Rubidium is geochemically associated with lithium, with higher concentrations in acidic igneous rocks and sedimentary aluminosilicates. In soil, it reacts similarly to potassium, and with weathering, the potassium/rubidium ratio continually decreases.

Plants

Its function is similar to that of potassium in some plants and it is easily taken up by plants. Although similar to potassium in availability and translocation in plants, rubidium does not substitute for potassium in biological functions

Animals/Man

Daily dietary intake is 1.5 to 6 mg.

Mobility in Food Chain

Rubidium has a high bioaccumulation index.

SELENIUM (Se)

Obtained as silvery metallic allotrope or red amorphous powder which is less stable; burns in air; unaffected by water; dissolves in concentrated nitric acid and alkalis

---

Atomic Number: 34   **Abundance in Lithosphere**: 0.05 mg/kg
Atomic Weight: 78.96   **Common Valence States**: Se^{6+}, Se^{4+}

Common Mineral Forms: traces in certain ores; obtained as a by-product in electrorefining of copper

---

Total Content in Soils: 0.1–2.0 mg/kg; when less than 0.6 mg/kg, plant content can be deficient in terms of animal health; average 0.40 mg/kg

Soluble Content in Soils: 0.02 mg/kg

Content in Sea Water: 1.8×10^{-7} to 0.47×10^{-7} mg/L

Content in Fresh Water: 0.6–20 μg/L; reference level 0.2 μg/L

Chemical Species in Water: SeO_3^{2-}, possible: $HSeO_3^-$, H_2SeO_3, SeO_4^{2-}, and $HSeO_4^-$

Content in Marine Animals: 4–5 mg/kg

Content in Animals: muscle, 0.05 mg/kg; kidney, 0.5 mg/kg

Content in Humans: muscle, 0.42–1.9 mg/kg; bone, 1–9 mg/kg; blood, 0.171 mg/dm^3

Maximum Daily Intake from Toy Materials: 5.0 μg

Content in Plants: reference plant, 0.02 mg/kg

Content in Phosphate Fertilizer: 0.3–2.1 mg/kg; rock phosphate, 0.2–11.0 mg/kg

Content in Common Foods: wheat grain, 0.15–0.5 mg/kg; agricultural raw product, 0.0016–0.37 μg/g wet weight

Essentiality: plants, no; animals, yes

General

The different oxidation forms of selenium will determine its soil and plant chemistry, which affects availability and potential toxicity, primarily in terms of animal health. In general, selenium increases with increase in clay content of soil. The solubility of selenium is relatively low, but on poorly drained or calcareous soils and arid soils when selenium soil levels are high, it may accumulate to sufficient levels in plants to pose a significant hazard to grazing animals. There are areas of the world (i.e., Finland and China) where selenium soil levels are not sufficient to assure that adequate selenium is introduced into the diet from locally consumed food/feed crops.

Plants

Selenium is not an essential element for plants, although it is being added to soils to ensure that both feed and food products contain sufficient amounts to meet the dietary needs of animals and man in selenium-deficient areas. Plants deficient in dietary selenium (less than 100 μg/kg) are found on soils containing less than 0.6 mg/kg. The solubility of selenium in soils is low, although soluble selenium is readily absorbed by plants. The availability of selenium is determined by climatic conditions (availability decreases with decreasing temperature and increasing rainfall), oxidation-reduction conditions, pH, and sesquioxide content of the soil. Some plant species are high accumulators of selenium, which can pose a danger to grazing animals if eaten. Selenium uptake and accumulation in plants are affected by other elements, such as nitrogen, phosphorus, and sulfur, as well as most of the micronutrients, including cadmium.

Animals

Muscle has the largest proportion of total body selenium, while the kidney has the highest selenium concentration. Hair selenium levels between 0.06 and 0.23 mg/kg in cows can result in white muscle disease in newborn calves. The maximum tolerable dietary level has been proposed as 2 mg/kg. Consumption of feedstuffs containing both toxic (>5 mg/kg) and deficient (<0.1 mg/kg) levels of selenium poses worldwide problems for livestock, particularly grazing animals.

Man

Daily dietary intake is 0.006 to 0.2 mg, toxic intake is 5 mg, and total mass of the element in an average (70-kg) person is 7 mg. The minimum daily requirement for selenium is 30 μg, but a safe intake is 20 to 70 μg/day.

Mobility in Food Chain

The index for bioaccumulation is 1 for selenium.

SILICON (Si)

Black amorphous silicon obtained by reduction of sand (SiO_2) with carbon; ultrapure semiconductor-grade crystals are blue-grey metallic; bulk Si unreactive toward oxygen, water, and acid (except HF), but dissolves in hot alkali

Atomic Number: 14 **Abundance in Lithosphere**: 277,000 mg/kg
Atomic Weight: 28.0855 **Common Valence State**: Si^{4+}

Common Mineral Forms: quartz (SiO_2); many and varied silicates

Total Content in Soils: 300 g/kg mostly as SiO_2 or silicates

Soluble Content in Soils: deficient when less than 20 mg/kg (in 0.1 M HOAc); 1–200 mg/L mainly as H_4SiO_4

Content in Sea Water: 0.03–4.09 mg/L

Content in Fresh Water: 6.5 mg/L

Content in Marine Animals: 70–1000 mg/kg (accumulated by Foraminiferae and some molluscs)

Content in Humans: muscle, 100–200 mg/kg; bone, 17 mg/kg; blood, 3.9 mg/dm^3; milk, 0.47 mg/L

Content in Animals: 120–6000 mg/kg; milk, 0.16–0.58 mg/L

Content in Plants: 200–5000 mg/kg

Essentiality: plants, no (beneficial); animals, no

General

Silicon is the most abundant and stable element in the lithosphere, but under specific conditions, it will dissolve and be transported. All silicate minerals contain tetrahedron SiO_4. Quartz (SiO_2) is the most resistant mineral in soils, while amorphous forms of silicate contribute to anion exchange capacity. Solubility of silicon varies with pH (more soluble in alkaline conditions), but the presence of other elements (aluminum, calcium, iron, and phosphorus) affects solubility as well as the presence of organic matter. Opals, a non-crystalline form of silicon found in soils, are derived from biological sources and have been examined as a means of identifying the presence of past plant species.

Plants

Silicon has been found to be beneficial to plants, accounting for stalk strength in grain crops (particularly rice), as well as providing some resistance to fungus and insect infestations and reducing water loss. Silicon also provides some resistance to manganese toxicity and may be related to phosphorus, aluminum, and heavy metal nutrition of plants. If silicon is required by plants, its critical value must be less than 6 mg/kg of dry tissue. Plants that accumulate silicon may contain as much as 4%, while those that exclude silicon (dicotyledons) contain less than 0.5%. Therefore, the silicon content in plants can vary widely. Silicon may be passively taken up by the plant, influenced by water flow, may be actively absorbed, or may be prevented from uptake in sizeable amounts by precipitation on root surfaces or binding within the root itself. For an accurate determination of silicon in plant tissue, care must be taken to ensure that tissue is free of dust or soil particles.

Animals/Man

Silicon is required for skeletal development of rats and chicks; the requirement is between 50 to 500 mg/kg. Lack of sufficient silicon may also be linked with several human disorders. Inhalation of high levels of silicon-bearing dust results in silicosis, a chronic lung disease. Consumption of high-silicon plants by ruminant animals may result in a disorder called siliceous renal calculi. The presence of opal phytoliths in pasture plants can result in considerable wear on sheep's teeth, since opals are harder than dental tissues. Silicon as quartz poses no health hazard.

Mobility in Food Chain

Not known.

SILVER (Ag)

Soft, malleable metal with characteristic silver sheen; stable to water and oxygen but attacked by sulfur compounds in air to form black sulfide layer; dissolves in sulfuric and nitric acid

Atomic Number: 47   **Abundance in Lithosphere**: 7.5×10^{-2} mg/kg
Atomic Weight: 107.868   **Common Valence States**: Ag^+, Ag^{2+}

Common Mineral Forms: argentite (silver glance) (Ag_2S); also obtained as a by-product of other metals such as copper

Total Content in Soils: 0.03–0.9 mg/kg

Soluble Content in Soils: 0.01–0.05 mg/kg in 1 N NH_4AOc

Content in Sea Water: 4×10^{-5} mg/kg

Content in Fresh Water: 0.13 µg/L

Content in Marine Animals: 3–10 mg/kg

Content in Humans: blood, <2.7 µg/L; bone, 1.1 mg/kg; liver, <5–32 ng/g

Content in Animals: 6 µg/kg

Content in Plants: 0.01–0.5 mg/kg

Content in Common Foods: 0.07–2.0 mg/kg

Essentiality: plants, no; animals, no

General

Silver has geochemical properties similar to copper, although it exists in the lithosphere at 1/1000th the level of copper. Silver is immobile in soil at pH greater than 4.0 and will be complexed by humic substances.

Plants

Silver content in plant tissue is usually less than 0.01 mg/kg, although in some soils, plant content may reach 0.5 mg/kg. Silver is quite toxic to plants in nutrient solution; 0.5 mg/L in solution results in a critical silver concentration in plants. Silver is considered one of the most toxic of the heavy metals to microorganisms.

Animals/Man

Daily dietary intake is 0.0014 to 0.08 mg, toxic intake is 60 mg, and lethal intake is 1.3 to 6.2 g. It is a suspected carcinogen. Silver in drinking water at 0.05 mg/L is considered the maximum contaminant level. For swine and poultry, the maximum tolerable level of dietary silver is 100 mg/kg.

Mobility in Food Chain

Silver has a moderate bioaccumulation index.

STRONTIUM (Sr)

Silvery white, soft metal; protected as bulk by oxide film but will burn in air and reacts with water

---

Atomic Number: 38   **Abundance in Lithosphere**: 370 mg/kg
Atomic Weight: 87.62   **Common Valence State**: Sr^{2+}

Common Mineral Forms: celestite ($SrSO_4$), strontianite ($SrCO_3$)

Total Content in Soils: 50–1000 mg/kg (average 150 mg/kg)

Soluble Content in Soils: 0.5–32 mg/kg in 1 N NH_4AOc

Content in Sea Water: 7.6 mg/L

Content in Fresh Water: reference level 50 µg/L

Chemical Species in Water: Sr^{2+}, possibly $SrOH^+$

Content in Humans: muscle, 0.12–0.35 mg/kg; bone, 36–100 mg/kg; blood, 0.031 mg/dm^3

Content in Animals: 14 mg/kg

Content in Plants: 3–3000 mg/kg; reference plant, 50 mg/kg

Essentiality: plants, no; animals, no

General

Strontium is a common trace element in the lithosphere. Its geochemical and biochemical characteristics are similar to calcium and to a lesser degree magnesium. The strontium/calcium ratio is relatively stable in the biosphere. Strontium is easily mobilized and its content in soils is controlled by the parent material and climate. The radioactive form of strontium, ^{90}Sr, was of considerable concern when atomic weapons were being tested in the atmosphere, as this radioactive element is considered one of the most biochemically hazardous to man.

Plants

The strontium content of plant tissue varies widely among plant species, from as low as 1 mg/kg to concentrations as high as 15,000 mg/kg. The strontium content of grains is very low, while that in legume herbage can be quite high. Uptake of strontium is fairly complex and is influenced to a degree by calcium.

Animals/Man

Daily dietary intake is 0.8 to 5 mg, and total mass of the element in an average (70-kg) person is 320 mg. Strontium is non-toxic. Since it readily accumulates in bone tissue, the addition of ^{90}Sr into the environment from atmospheric atomic weapon testing is considered a major health threat to man.

Mobility in Food Chain

Strontium has a moderate bioaccumulation index.

THALLIUM (TI)

Soft, silvery-gray metal; tarnishes readily in moist air and with steam reacts to form TlOH; attacked by acids, rapidly by HNO_3

Atomic Number: 81   **Abundance in Lithosphere**: 0.6 mg/kg
Atomic Weight: 204.3833   **Common Valence State**: Tl^+ (rarely Tl^{3+})

Common Mineral Forms: lorandite ($TlAsS_2$), jarosite (K_2SO_4; $3FeSO_6 \cdot 6H_2O$) contains relatively high contents of Tl; granitic-rhyolitic rocks may contain 3 mg/kg Tl

Total Content in Soils: 0.2–2.8 mg/kg; mean value 0.2 mg/kg

Soluble Content in Soils: 10–20% of the total Tl content

Content in Sea Water: 1.4×10^{-5} mg/kg

Content in Fresh Water: 0.03 μg/L

Content in Human Tissues: blood, 0.5 μg/L; bone, 2 ng/g; liver, 0.56–2.85 ng/g

Content in Animals: 100 μg/kg

Content in Plants: 0.02–0.125 mg/kg

Content in Fertilizers: potassium containing, 0.1 mg/kg

Essentiality: plants, no; animals, no

General

Thallium is widely distributed in the lithosphere and enters the environment primarily from coal burning and cement manufacture. Thallium can be complexed by organic matter and can be methylated, forming volatile compounds. The thallium content in soil is typically near 0.5 mg/kg, and its uptake by plants is similar to that of potassium. Thallium is easily mobilized in soil, as it is transported together with the alkaline metals.

Plants

The uptake of thallium by plants closely parallels the thallium content of the soil. In high thallium content soils, plants can have unusually high thallium levels, sometimes as high as 10 mg/kg. Tobacco is relatively sensitive to thallium. High thallium hinders seed germination and affects both photosynthesis and transpiration. Thallium toxicity results in visual leaf chlorosis. Kohlrabi is a thallium accumulator plant. Thallium is detectable in some wines, usually red (0.2 μg/L) more than white, and is found to be highest in the flesh of the grape berry and lowest in the stem of bunches. Thallium can accumulate to concentrations as high as 100 mg/kg in some plants. Concentrations in excess of 20 mg/kg are considered toxic. Soil application of thallium as thallium sulfate at 150 mg/kg was found to be lethal to collard and wheat plants.

Animals/Man

The average human intake is estimated to be 1.5 to 2.0 µg/day; daily urinary excretion ranges from 0.02 to 9.0 µg/L. Thallium is moderately toxic and accumulates in different organs with age. Tachycardia and hypertension are the most common signs of thallium toxicity. Thallium may activate some enzymes, like pyruvate kinase. Calcium can interfere with the intake of thallium. Lethal intake is 600 mg. Accumulation of thallium can lead to automatic dysfunction, with tachycardia and hypertension.

Mobility in Food Chain

Thallium has a very low bioaccumulation index.

TIN (Sn)

Soft, pliable, silvery-white metal; unreactive in oxygen (protected by oxide film) and water but dissolves in acids and bases

Atomic Number: 50	**Abundance in Lithosphere**: 40 mg/kg
Atomic Weight: 118.69	**Common Valence State**: Sn^{2+}, Sn^{4+}

Common Mineral Forms: cassiterite (SnO_2)

Total Content in Soils: 0.3–200 mg/kg; mean value 1.1 mg/kg (in peats, 50–300 mg/kg)

Soluble Content in Soils: less than 1 mg/kg

Content in Sea Water: 4 ng/L; range 0.01–0.3 ng/g

Content in Fresh Water: river water, 0.3–17 ng/g; reference level 0.01 µg/L

Chemical Species in Water: mono-, di-, and trimethyl compounds

Content in Humans: muscle, 0.33–2.4 mg/kg; bone, 1.4 mg/kg; blood, 0.38 mg/dm^3

Content in Animals: 0.1–0.9 mg/kg

Content in Plants: 0.1–3.0 mg/kg (ferns, 30 mg/kg); reference plant, 0.2 mg/kg

Content in Common Foods: 0.04–0.1 mg/kg

Content in Fertilizers: 0.2 mg/kg (range 10–15 mg/kg)

Essentiality: plants, no; animals, no

General

The mobility of tin in soil is pH dependent, similar to iron and aluminum. Tin can be complexed by soil organic matter and therefore can be accumulated in soils high in organic matter content. Tin is used as a coating on metal food cans, but its release into the contained food product is minimal.

Plants

Tin can be easily taken up by plants from a nutrient solution but less so in acid soils, although it primarily accumulates in the roots and is not easily translocated up the plant. Reported tin contents vary widely, from less than 1 mg/kg to as high as 300 mg/kg. The normal range seems to be between 5 to 10 mg/kg, with sedges and mosses being high accumulators. There is no evidence that tin is toxic to plants unless at high concentrations.

Animals/Man

Dietary intake is 5.8 mg/day, toxic intake is 2 g (most organotin compounds are very toxic), and total mass of the element in an average (70-kg) person is 14 mg. Tin can be deficient when at very low levels in animal diets but toxic at higher levels of intake, particularly organic tin compounds. Inorganic tin compounds are not easily absorbed.

Mobility in Food Chain

Not known.

TITANIUM (Ti)

Hard, lustrous metal; resists corrosion due to oxide layer, but powdered metal burns in air; unaffected by many acids (except HF, H_3PO_4, and concentrated H_2SO_4), and alkalis

---

Atomic Number: 22   **Abundance in Lithosphere**: 4400 mg/kg
Atomic Weight: 47.88   **Common Valence State**: Ti^{4+}

Common Mineral Forms: rutile (anatase, boolite) as TiO_2, ilmenite ($FeTiO_3$), perowskite ($CaTiO_3$), titantishere ($CaTiO_3$-$CaTiSiO_5$)

Total Content in Soils: 1800–3600 mg/kg; mean value 2400 mg/kg

Soluble Content in Soils: less than 1 mg/kg

Content in Sea Water: 1 µg/L (4.8×10^{-4} mg/L)

Content in Fresh Water: 8 ng/L

Content in Marine Animals: 20 µg/kg; diatoms, 15–15,000 mg/kg

Content in Man: muscle, 0.9–2.2 mg/kg; blood, 0.0054 mg/dm^3

Content in Animals: 0.1–0.9 mg/kg

Content in Plants: 0.1–4.6 mg/kg

Content in Common Foods: 0.5–6.1 mg/kg

Content in Fertilizers: 600 mg/kg (range 100–3000 mg/kg)

Essentiality: plants, no; animals, no

General

Titanium is the ninth most abundant element in the lithosphere. Most of its forms are very insoluble in water; therefore, it accumulates in surface soils. The common oxidation number for titanium is 4+, although under special redox potentials it may exist in the 3+ form and rarely in the 2+ form. Titanium as a metal is generally considered physiologically inactive in plants, animals, and man due to its very low bioavailability. Due to its low absorption and retention, levels in plant and animal tissues are lower than would be expected due to the high abundance of titanium in the lithosphere.

Plants

Titanium is found in low concentrations in plants, generally less than 1 mg/ kg, although leaf concentrations as high as 80 mg/kg have been reported. Since titanium is a major constituent in soil and dust, care must be taken to adequately decontaminate plant tissue prior to an assay in order to obtain an accurate titanium value. A titanium determination has been suggested by some as an indicator of soil/dust contamination. Although not considered an essential element for plants, some reports suggest that titanium may possibly participate in photosynthesis and molecular nitrogen (N_2) fixation processes and therefore be beneficial to plants. When titanium exceeds 200 mg/ kg in plant tissues, chlorosis and necrotic spots will appear on plant leaves. In a nutrient solution, 4.5 mg/kg titanium in solution has been shown to be toxic to bush beans, with most of the titanium accumulating in the roots.

Animals/Man

The titanium content of various body parts is highly variable. The lungs are fairly high in titanium (>4 µg/g fresh tissue) due primarily to dust ingestion. Mean titanium content of various human organs is as follows: muscle, 0.2 ± 0.01; brain, 0.8 ± 0.05; kidney cortex, 1.3 ± 0.2; kidney medulla, 1.2 ± 0.2; liver, 1.3 ± 0.2; and lungs, 3.7 ± 0.9 µg/g fresh tissue. Mean contents of 10.6 ± 19.3 µg/g in hair, 0.46 ± 0.13 µg/g in tooth enamel, and 0.07 µg/ g in whole blood have been reported. Titanium is considered essentially non-toxic in the amounts and forms that are usually ingested. Determination of the titanium content of feces from small children can be used as an indicator of the degree of soil ingestion.

Chelated Forms

Recent research has shown that a chelated form of titanium, titanium ascorbate (trade name TITAVIT), can have beneficial effects on plants, animals, and man.

Mobility in Food Chain

Titanium is moderately mobile within the food chain.

TUNGSTEN (W)

Lustrous and silvery white metal; resists attack by oxygen, acids, and alkalis

---

Atomic Number: 74   **Abundance in Lithosphere**: 1.0 mg/kg
Atomic Weight: 183.85   **Common Valence States**: W^{4+}, W^{6+}

Common Mineral Forms: scheelite ($CaWO_4$), wolframite [$(Fe,Mn)WO_4$]

Total Content in Soils: 1.5 mg/kg

Content in Sea Water: 9.2×10^{-5} mg/L

Content in Human Body: bone, 0.00025 mg/kg; blood, 0.001 mg/dm^3

Content in Animals: 1–2 µg/kg

Content in Plants: 0.0–0.15 mg/kg; reference plant, 0.2 mg/kg

Content in Phosphate Fertilizers: 2.3 mg/kg

Content in Rock Phosphate: 1.4–2.7 mg/kg

Content in Food Plants: <0.001–0.35 mg/kg; 0.02–0.13 mg/kg in edible parts of vegetables

Essentiality: plants, no; animals, no

General

Tungsten is not evenly distributed, as it is affected by acidity of the parent material and soil-forming processes; tungsten content is thus higher under acidic conditions. Its geochemical behavior is similar to that of molybdenum.

Plants

Little is known about this element in terms of its function in plants, whether its presence has either a stimulatory or toxic effect. Some have suggested that tungsten can substitute for molybdenum in the nitrate reductase enzyme. It has been observed that *Pinus sibiricus* is an accumulator plant of tungsten and that tungsten can be readily taken up by plants under certain soil conditions.

Animals/Man

Little is known about this element in terms of its animal or human physiology. In animal studies, lack of tungsten was found to increase the mortality rate of goats. Most ingested tungsten is excreted in urine and feces and is not retained in the body.

Mobility in Food Chain

Not known.

URANIUM (U)

Radioactive silvery metal; malleable, ductile, and tarnishes in air; attacked by steam and acids but not alkalis

Atomic Number: 92   **Abundance in Lithosphere**: 2.4 mg/kg
Atomic Weight: 238.0289   **Common Valence State**: U^{3+}

Common Mineral Forms: uraninite (U_3O_8), carnotite [$K_2(U)_2(VO_4)_2 \cdot 2H_2O$]

Total Content in Soils: 0.10–11.2 mg/kg (average range 0.79–3.70 mg/kg)

Content in Fresh Water: 0.05 µg/L

Chemical Species in Water: UO_2^{2+}, $UO_2(CO_3)_3^{4-}$, $UO_2(CO_2^{2-}$, $UO_2(HPO_4)_2^{2-}$, $UO_2(CO_3)_3^{4-}$

Content in Sea Water: 3.13 µg/L

Content in Human Body: muscle, 0.9 µg/kg; bone, 0.016–70 µg/kg; blood, 5×10^{-4} mg/dm^3

Content in Animals: 13 µg/kg

Content in Plants: 0.5–60 µg/kg; reference plant, 0.01 mg/kg

Content in Fertilizers: rock phosphate, 120 mg/kg

Content in Common Foods: corn and potato, 0.8 µg/kg; other foods, 2 µg/kg

Essentiality: plants, no; animals, no

General

Uranium is found at higher concentrations in acid rocks than mafic rocks and sediments, and its biochemical and soil chemistry are somewhat complex. Uranium's chemistry in the environment has become important due to its atomic characteristics and use.

Plants

Background values are in the range <1 to 6 mg/kg. Excess uranium in plants can be toxic, affecting chromosome numbers and producing unusually shaped fruit and leaf rosettes. Uranium will precipitate in the root tips as autunite [$Ca(UO_2)_2PO_4$].

Animals/Man

Daily dietary intake is 0.001 to 0.002 mg, lethal intake is 36 mg (rats), and total mass of the element in an average (70-kg) person is 0.09 mg.

Mobility in Food Chain

Not known.

VANADIUM (V)

Shiny, silvery metal, soft when pure; resists corrosion due to protective oxide film; attacked by concentrated acids but not by fused alkalis

---

Atomic Number: 23   **Abundance in Lithosphere**: 160 mg/kg

Atomic Weight: 50.9415   **Common Valence States**: V^{3+}, V^{4+}, V^{5+} (very rare: V^{2+})

Common Mineral Forms: vanadinate: $Pb_5(VO_4)_3$, patronite (VS_4), pucherite ($BiVO_4$); different bauxites contain sometimes high vanadium content (0.1% V_2O_5) and some crude oils have a vanadium content from 1–2 kg/ton

---

Total Content in Soils: 3–230 mg/kg (average 90 mg/kg)

Content in Sea Water: 2.4 µg/L

Content in Fresh Water: 0.3–20 µg/L; reference level 1.0 µg/L

Chemical Species in Water: $H_2VO_4^-$ or HVO_4^{2-}

Content in Marine Animals: 1 mg/kg; can accumulate to very high levels (25–300 ng/g wet weight) in crustaceans, shellfish, and fish

Content in Human Body: blood, <0.1 µg/L; bone, 0.8–8.3 ng/g; liver, 2.53–13.4 ng/g; total mass, 0.11 mg

Content in Animals: 0.1 mg/kg

Content in Plants: 0.27–4.2 mg/kg (average 1 mg/kg); reference plant, 0.5 mg/kg

Content in Fertilizers: 20–500 mg/kg

Content in Common Foods: lettuce, 2 mg/kg; milk, 6 µg/L; range 0.5–760 µg/kg

Essentiality: plants, no (may be essential for certain algae and bacteria); animals, no (yes for chick and rat)

General

There is wide variation in the vanadium content of rocks, and its geochemical properties are highly dependent on its oxidation state and the acidity of the media. Vanadium is associated with clay minerals, iron oxides, and organic matter and can be quite high in organic shales and bioliths. The anionic forms of vanadium (VO_4^{3-} and VO_3^-) are mobile in soils and can be relatively toxic to soil microbiota. Vanadium deposition from air pollution can be significant around smelters and near coal- and oil-burning plants, with soil levels of 100 mg/kg or greater.

Plants

Vanadium at 0.5 mg/kg in a nutrient solution has been found to be toxic to plants. Mature corn leaves were found to contain 0.36 to 1.05 mg/kg. Vanadium at 2 mg/kg or more in pea and soybean plants was found to be toxic. There is wide variation in vanadium content (<5 to 50 mg/kg) found in plants. Some of this variation is due to air pollution when the vanadium content exceeds 100 mg/kg; therefore, there is a need to distinguish between that absorbed through the roots and transported to the tops and that received by aerial deposition. There is some evidence that vanadium may be involved in nitrogen (N_2) fixation and may partially substitute for molybdenum in various biological nitrogen transformations; the level required is 2 µg/kg. Vanadium is readily taken up through plant roots, and uptake is strongly influenced by the ionic form of vanadium and pH (absorption decreases with increasing pH). Some plants can accumulate substantial quantities of vanadium.

Animals/Man

Daily dietary intake is 0.04 mg. Vanadium concentration in the human body is very low, from 0.01 to 0.6 µg/g wet weight, with the lungs having higher vanadium content due to air pollution. Vanadium is not normally toxic when ingested but can be toxic when entering by way of the respiratory system. Vanadium levels in air as low as 0.01 to 0.04 m/m³ are potentially toxic. Diets containing less than 10 µg/kg are deficient for the normal growth of chicks and rats.

Mobility in Food Chain

Vanadium has a very low bioaccumulation index.

ZINC (Zn)

Bluish-white metal, brittle when cast; tarnishes in air, reacts with acids and alkalis

Atomic Number: 30	**Abundance in Lithosphere**: 80 mg/kg
Atomic Weight: 65.39	**Common Valence State**: Zn^{2+}

Common Mineral Forms: zinc blende (ZnS), calamine (smithsonite) ($ZnCO_3$), and sphalerite [(Zn,Fe)S]

Total Content in Soils: 10–300 mg/kg

Soluble Content in Soils: 4–270 µg/L

Content in Sea Water: $0.5–1.0 \times 10^{-4}$ mg/L

Content in Fresh Water: 0.1–240 µg/L; reference level 5.0 µg/L

Chemical Species in Fresh Water: Zn^{2+}

Content in Marine Animals: 6–1500 mg/kg (will accumulate in Radiolaria and Molluscae)

Content in Animals: 160 mg/kg; milk, 1.8–4.2 mg/L

Content in Humans: muscle, 240 mg/kg; bone, 75–170 mg/kg; blood, 7.0 mg/dm^3; milk, 2.15 mg/L

Content in Plants: 10–100 mg/kg (critical value 15 mg/kg); reference plant, 50 mg/kg

Content in Common Foods: 1–150 mg/kg; raw agricultural products, 1.5–42.0 µg/g wet weight

Essentiality: plants, yes; animals, yes

General

Zinc is fairly uniformly distributed and in soils is easily adsorbed by mineral and organic substances, accumulating primarily in the soil surface layer. Zinc soil chemistry is quite complex in terms of its various combined ionic forms, with zinc being the most readily soluble of all the heavy metals in soils. Zinc plant availability decreases with increasing soil pH.

Plants

Zinc is an essential micronutrient with a sufficiency range of 20 to 100 mg/kg dry weight of tissue. The critical zinc value has been accepted as 15 mg/kg for most crops, although under some conditions 10 mg/kg may be sufficient. Zinc tends to accumulate in older leaves. A typical symptom of zinc deficiency is rosetting of terminals. Zinc deficiency is most likely to occur on leached sandy soils low in organic matter content, neutral to alkaline soils, and/or soils high in available phosphorus. The effect of phosphorus on zinc is thought to occur within the plant and is not a major factor in zinc uptake. Other important interactions with copper, iron, arsenic, and nitrogen affect zinc metabolism. A number of major food crops, such as corn, sorghum, citrus, and field bean, are particularly sensitive to zinc. Toxicity can occur when soil zinc levels are elevated by the application of industrial wastes or sewage sludge, which can contain zinc levels in excess of 1700 mg/kg. The effects of zinc toxicity can be significantly reduced by liming the soil to pH 6.0 or above.

Animals/Man

Daily dietary intake is 5 to 40 mg, toxic intake is 150 to 600 mg, lethal intake is 6 g, and total mass of the element in an average (70-kg) person is 2.3 g. Zinc is widely distributed throughout the body, although regular daily intake is essential since the body does not store zinc. Many common dietary components (phytic acid) can significantly reduce the absorption of zinc, resulting in a zinc deficiency even when food/feedstuffs may contain adequate zinc for the diet. Zinc is well known regarding its relationship to skin and wound healing.

Mobility in Food Chain

Zinc has a fairly high bioaccumulation index.

ZIRCONIUM (Zr)

Hard, lustrous, silvery metal; very corrosion resistant due to oxide layer, but will burn in air; unaffected by acids (except HF) and alkalis

---

Atomic Number: 40   **Abundance in Lithosphere**: 165 mg/kg
Atomic Weight: 91.22   **Common Valence State**: Zr^{4+}

Common Mineral Forms: zircone [$Zr(Hf)SiO_4$], baddelayite (ZrO_2, very rare); 400–500 mg/kg in some bauxites

Total Content in Soils: mean ~250 mg/kg (range 30–2000 mg/kg)

Content in Sea Water: 4 µg/L

Content in Fresh Water: 2.6 µg/L

Content in Humans: muscle, 0.08 mg/kg; blood, 0.011 mg/dm^3

Content in Plants: 0.3–2.0 (if higher, due to soil pollution); mean 0.6 mg/kg; reference plant, 0.1 mg/kg

Content in Animals: less than 0.3 µg/kg

Content in Foods: tea is rich in Zr (12 µg/g)

Content in Fertilizers: 30 mg/kg (range 10–800 mg/kg); phosphate, 50 mg/kg

Essentiality: plants, no; animals, no

General

Zirconium is the twelfth most common element in the earth's crust, and content is consistent among soil types. Our knowledge of the physiological roles of zirconium is limited because the solubility of its compounds under common pH values is very low; therefore, its availability to living systems is low. The common oxidation number is +4, although the +3 form can exist in special circumstances and the +2 form in very rare instances. Zirconium is considered to be physiologically inactive in plants, animals, and man. The common form of zirconium for uptake is as the anion $Zr(OH)_n^{4-n}$.

Plants

Zirconium is found in low concentrations in plants, generally less than 1 mg/kg; variation in food plant content is 0.005 to 2.6 mg/kg. Although not considered an essential element for plants, some reports suggest that zirconium is similar to titanium in that it participates in various physiological processes. Chlorella green-algae growth was stimulated if applied zirconium ascorbate was present at 0.1 to 1.0 µ*M*. The zirconium treatment can change the ratio of different photosynthetic pigments, with the algal cells showing a definite accumulation of zirconium. In yeast experiments using zirconium ascorbate or zirconium citrate, this element does not inhibit cell production of *Saccharomyces cerevisiae* or *Candida utilis*. It is interesting that zirconium treatment enhanced protein synthesis in general and markedly changed the amino acid composition of the proteins. In this respect, zirconium acts very similar to titanium. It has also been demonstrated that zirconium can reduce the phyto-availability of phosphates for phytoplanktons. Zirconium in excess of 15 mg/kg is considered toxic.

Animals/Man

Average human intake is 0.05 mg/day. Concentration in different human tissues ranges between 10 to 60 µg/kg fresh weight. Literature values for blood are between 10 and 30 µg/L. Zirconium is not considered toxic, but data on its potential carcinogenicity or teratogenicity exist.

Chelated Forms

Similar to titanium, zirconium forms different organic chelates with ascorbic acid, citric acid, etc. These forms are water soluble and pH stable. According to recent literature, there may be a possible beneficial physiological role for zirconium in such chelated forms.

Mobility in Food Chain

Zirconium has a very low bioaccumulation index.

3

RARE EARTH ELEMENTS

This group of elements, also known as *lanthanides*, consists of 14 elements, not including promethium, which is an artificial element. The group derived its name because these elements are not concentrated in ores, like most metals; therefore, their minerals very seldom exist. Their total concentration in the lithosphere is about 200 mg/kg, and some rare earth elements, such as cerium, neodymium, and lanthanum, are more frequently associated with some physiological role than are a number of the known essential elements. The associations are well known and documented.

Atomic number and weight, abundance in the lithosphere, and mean level in soil for the rare earths are given in Table 3.1

There are some common mineral forms for several of the rare earths: monacite ($CePO_4$) which also contains lanthanum, ytterbium, erbium, neodymium, and praseodymium; xenotine ($YbPO_4$); and lanthanite $La(Nd,Pr,Ce)_2(CO_3)_2$.

Monacite sand is used worldwide for industrial purposes. In 100 kg of this material, the content of rare earth elements is:

600 g gadolinium (Gd)	70 g ytterbium (Yb)	11.0 kg lanthanum (La)
220 g dysprosium (Dy)	45 g europium (Eu)	9.0 kg neodymium (Nd)
170 g erbium (Er)	45 g terbium (Tb)	2.7 kg praseodymium (Pr)
150 g thullium (Tm)	5 g lutetium (Lu)	1.0 kg samarium (Sm)
130 g holmium (Ho)	21.0 kg cerium (Ce)	

TABLE 3.1   Rare earth elements

Element (Symbol)	Atomic Number	Atomic Weight	mg/kg	
			Abundance in Lithosphere	Mean Level in Soil
Lanthanum (La)	57	138.92	30.0	41.2
Cerium (Ce)	58	140.13	60.0	84.2
Praseodymium (Pr)	59	140.92	8.2	6.5
Neodymium (Nd)	60	144.27	28.0	43.6
Samarium (Sm)	62	150.43	6.0	6.0
Europium (Eu)	63	152.00	1.2	1.3
Gadolinium (Gd)	64	156.90	5.4	3.5
Terbium (Tb)	65	158.93	0.9	0.8
Dysprosium (Dy)	66	162.46	3.0	5.6
Holmium (Ho)	67	164.94	1.2	0.8
Erbium (Er)	68	167.20	2.8	3.0
Thullium (Tm)	69	168.94	0.5	0.6
Ytterbium (Yb)	70	173.04	3.0	3.9
Lutetium (Lu)	71	174.99	0.5	0.4

China is where much of the soluble and easily absorbed rare earth elements are found. The highest values have been found in southeast China, where the laterite red earth, red earth, paddy, and terra rossa soil types have the highest availability. The common parent material from which these soils were formed is granite, which is the source of the rare earth elements.

Although much of the interest in the rare earth elements has occurred in China (Guo, 1985, 1987; Zhu and Liu, 1991), an evaluation of their published literature poses two difficulties: (1) only a few papers have been published in English and (2) the research usually deals with water-soluble mixtures of the different rare earth elements, which makes identification of an observed physiological effect difficult to attribute to only one of the elements (Brown et al., 1990).

Other than their mean level in soil, the bioavailability of the rare earth elements is not very well known; consequently, neither are their physiological roles. However, these elements have been studied to a limited degree, and some of the known information about them in the literature is summarized here, selecting only data that could be accepted based on verified results.

The average human body uptake of these elements is estimated at ~2 mg/day. Known data on urinary excretion is as follows:

Europium (Eu)	3–360 ng/L
Gadolinium (Gd)	<1 μg/L
Lutetium (Lu)	1–200 ng/L
Samarium (Sm)	1–210 ng/L
Ytterbium (Yb)	5–90 ng/L

PLANTS

The effects of the rare earths on plant growth have been reported by Alejar et al. (1988), Meehan et al. (1993), and Diatloff et al. (1995).

Anabaena asotica, a cyanobacteria, has increased nitrogen fixation activity in the presence of lanthanum chloride ($LaCl_3$) as well as praseodymium chloride ($PrCl_3$). In the presence of these two elements, plant growth is also promoted. Interestingly, the same authors (Wang et al., 1985) found that $CeCl_3$ is an inhibitor of nitrogen fixation.

Several NBS Standard Reference Materials (National Institute of Standards and Technology, Gaithersburg, MD) have been analyzed by the neutron activation method, with reported values for several rare earth elements. Meloni and Genova (1987) found concentrations between 10 and 1500 ng/g. The highest and lowest values were found in spinach (RSM 1570a) and orchard leaves (no longer available): 365 and 1440 ng lanthanum, 765 and 1280 ng cerium, 3.5 and 5.4 ng lutecium, and 4.9 and 7.0 ng thullium. In bovine liver (RSM 1577b), they found 17 ng lanthanum, 48 ng cerium, and 0.1 ng lutecium and thullium. It is interesting to note that the plant tissue standards have much higher contents of the rare earth metals than does the one animal tissue, bovine liver.

In the Liu (1988) publication, there are some data on the rare earth content of different foods: wheat, 1.0; corn, 0.2; rice, 0.6; and tomato, 0.05 μg/g. The highest rare earth content was 3.61 μg/g, which was found in sesame.

Amann et al. (1992) found that europium ions can activate the NAD-kinase enzyme in plants. This ion can replace calcium in calmodulin, which explains this promotional effect.

Different fungi, *Aspergillus niger* and *Mucor rouxii,* can accumulate lanthanum by using phenolic compounds. A similar accumulation was demonstrated by Ichihashi et al. (1992) in the case of the herb *Phytolacea americana* and two ferns; the accumulation rate was related to the rare earth content of the soil for the elements lanthanum, ytterbium, and dysprosium, the elements with the highest content in the leaves.

Summerton (1992) found a high accumulation rate for the lighter rare earth elements in a salt-tolerant plant, samphire, which shows significant discrimination in uptake of the rare earth elements.

ANIMALS

Using the neutron activation analysis technique, Arambel et al. (1986) determined the cerium, lanthanum, and samarium content of duodenal digests which were collected from cattle fed a diet containing rare earth elements. In a similar paper, it was shown that ytterbium-labeled forage appears to be more effective as a particulate marker for measurements made on duodenal digests (Andersen et al., 1985).

Collys et al. (1992) demonstrated that lanthanum can increase the fluorine content in bovine enamel *in vitro* from a calcifying solution.

Recently, an interaction was demonstrated between calf thymus DNA and several rare earth elements, such as lanthanum (La^{3+}), europium (Eu^{3+}), and terbium (Tb^{3+}). Comparison with the copper-DNA complexes shows some degree of helical destabilization of the biopolymer in the presence of lanthanide cations.

MAN

Trace metal content of various tissues of workers affected by rare earth pneumoconiosis was studied using the neutron activation analysis analytical method (Pietra et al., 1985). These workers were found to have very high concentrations of the rare earth elements in their lungs, and an increase in the neodymium and cerium content of their urine was also found. There was a magnitude of 4 increase in the cerium content of their lungs; the increase for lanthanum and samarium was 2000 to 3000 times.

In an article on inflammatory disease mechanisms where blood clotting plays an important role in the mechanism of inflammation, Jancsó (1961) applied lanthanum, cerium, neodymium, praseodymium, and samarium in the form of inorganic salts and in the second phase of his work synthetized new compounds of rare earth elements. Based on this research, which was conducted in Hungary and has also been done in other countries as well, the sulfosalicylic acid chelate of samarium was found to be an excellent anti-inflammatory agent

Based on cellular pharmacological research, lanthanum has been demonstrated as being a specific antagonist of calcium in biological systems. According to Weiss (1974), lanthanum can block all calcium movement by replacing the calcium ion on particular binding sites.

In other research conducted by Kádas and Jobst (1973) and Nagy et al. (1976), it was shown that higher amounts of lanthanum can produce pathological symptoms in the livers of rats and rabbits.

4

PLANT MICRONUTRIENTS

TERMINOLOGY

The seven micronutrients are boron, chlorine, copper, iron, manganese, molybdenum, and zinc. They have been variously identified in the past as either trace or minor elements, a term that is no longer used. The correct term is micronutrient.

Micronutrient concentrations are expressed in a variety of terms: parts per million (ppm) is most commonly used in the technical literature and SI units, either milligrams per kilogram (mg/kg), millimole per kilogram (mmol/kg), or equivalents, are used in the scientific literature. Comparative values for the micronutrients in these units are given in the following example:

Micronutrient	ppm[a]	mg/kg	mmol/kg	μequiv/100 g[b]
Boron (B)	20	20	1.85	554.6
Chlorine (Cl)	100	100	2.82	0.28
Copper (Cu)	12	12	0.19	37.7
Iron (Fe)	111	111	1.98	596.3
Manganese (Mn)	55	55	1.00	200.2

Micronutrient	ppm[a]	mg/kg	mmol/kg	μequiv/100 g[b]
Molybdenum (Mo)	1	1	0.01	—
Zinc (Zn)	33	33	0.50	100.9

[a] Parts per million (ppm) = milligrams per kilogram (mg/kg).
[b] mg/kg × factor = μequiv/100 g; factors: B = 27.726, Cl = 0.0028, Cu = 3.1476, Mn = 3.641, Zn = 3.059.

CONTENT IN PLANTS

The micronutrients are found and required in relatively low concentrations in plants as compared to the major elements. The approximate concentrations for the micronutrients in mature leaf tissue generalized for various plant species are given in Table 4.1.

TABLE 4.1   Approximate concentration of the micronutrients in mature leaf tissue generalized for various plant species

Micronutrient	mg/kg		
	Deficient	Sufficient or Normal	Excessive or Toxic
Boron (B)	5–30	10–200	50–200
Chlorine (Cl)	<100	100–500	500–1000
Copper (Cu)	2–5	5–30	20–100
Iron (Fe)	<50	100–500	>500
Manganese (Mn)	15–25	20–300	300–500
Molybdenum (Mo)	0.03–0.15	0.1–2.0	>100
Zinc (Zn)	1–20	27–100	100–400

Melsted (1973) has also given given normal ranges for micronutrient content in plant leaves, as well as maximums for plants in general and corn leaves in particular. These values are given in Table 4.2.

Plant contents for five of the micronutrients, including the range and common concentration for normal and toxic levels in plants, are given in Table 4.3.

TABLE 4.2   Normal range and suggested maximum micronutrient concentrations for plant leaves and suggested levels for corn leaves

| | Concentration (mg/kg dry weight) | | | |
| | Plant Leaves | | Corn Leaves | |
Element	Range	Maximum	Range	Maximum
Boron (B)	7–75	150	Same	100
Copper (Cu)	3–40	150	5–25	30
Iron (Fe)	20–300	750	50–200	300
Manganese (Mn)	15–150	300	Same	Same
Molybdenum (Mo)	0.2–1.0	3	0.2–1.0	5[a]
Zinc (Zn)	15–150	300	20–100	300

[a] The level of Mo which is toxic to cattle depends upon the concentration of Cu in rations. The recommended maximum could be toxic in very low Cu rations.

Source: Melsted, 1973.

More recently, Markert (1994a) published his listing of the trace element content of a reference plant, what might be considered normal for the trace element content of plants in general, and the values for the micronutrients, as given in Table 4.4.

TABLE 4.3   Total concentration of the micronutrients typically found in plants

| | μg/g | |
Micronutrient	Normal	Toxic[a]
Boron (B)	30–75	>75
Copper (Cu)	4–15	>20
Manganese (Mn)	15–100	—[b]
Molybdenum (Mo)	1.0–100	—
Zinc (Zn)	15–200	>200

[a] Toxicities listed do not apply to certain accumulator plants.
[b] — = no data.

TABLE 4.4 Micronutrient content of reference plant

Micronutrient	mg/kg
Boron (B)	40
Copper (Cu)	10
Iron (Fe)	150
Manganese (Mn)	200
Molybdenum (Mo)	0.5
Zinc (Zn)	50

Note: No data from typical accumulator and/or rejector plants.

Source: Markert, 1994a.

Typical micronutrient contents in plants in general, including the cell sap content, have been determined by Robb and Pierpoint (1983) for plants growing in a nutrient solution of specific micronutrient concentration. Those values for the micronutrients are given in Table 4.5.

The sufficiency concentration for the micronutrients in plants is necessary if an assay of plant content is to be used for diagnostic purposes (J.B. Jones, 1991). Such data are given for each micronutrient in this text; however, Bergmann (1983) and Bergmann and Neubert (1976) have set suffi-

TABLE 4.5 Typical concentrations of trace elements in foliage of normal plants

Trace Element	Dry Matter Content (mg/kg)	Cell Sap Content[a] (mM)	Nutrient Solution[b] (mM)
Boron (B)	15–100	0.2–1.3	0.05
Chlorine (Cl)	100–1000	0.4–4.0	0.1
Copper (Cl)	5–15	0.01–0.03	0.001
Iron (Fe)	50–300	0.15–0.75	0.1
Manganese (Mn)	25–250	0.06–0.6	0.01
Molybdenum (Mo)	0.5–5	0.004–0.075	0.0005
Zinc (Zn)	15–75	0.03–0.15	0.002

[a] This value merely indicates, in round numbers, the total aqueous concentration possible without regard to insoluble, lipid, or other structural fractions.
[b] Typical nutrient solution concentrations.

Source: Robb and Pierpoint, 1983.

ciency ranges for the micronutrients for 12 major agronomic crops that are grown throughout the world. Their sufficiency ranges are based on the Feeks 718 growth stage; the ranges are given in Table 4.6.

TABLE 4.6   Sufficient micronutrient content of plants at Feeks 718 growth stage

Crop	Boron (B)	Copper (Cu)	Iron (Fe)	Manganese (Mn)	Molybdenum (Mo)	Zinc (Zn)
				mg/kg		
Barley	5–10	5–10	21–200	25–150	0.10–0.3	15–60
Cotton	20–80	8–20	50–350	35–150	0.60–2.0	25–80
Groundnut	25–70	7–15	—	50–200	0.50–1.0	20–70
Maize	6–15	6–15	11–300	35–150	0.15–0.4	25–70
Oats	5–10	5–10	62–204	35–150	0.15–0.4	20–70
Potato	25–70	7–15	—	40–200	0.20–0.5	20–80
Rice	6–15	7–12	>80	40–150	0.40–1.0	30–70
Rye	4–10	5–10	—	20–100	0.10–0.3	15–60
Sorghum	5–15	5–12	50–250	25–150	0.15–0.3	15–60
Soybean	25–60	10–20	21–300	30–150	0.50–1.0	25–60
Sunflower	35–100	10–20	79–113	25–100	0.30–1.0	30–80
Wheat	5–10	5–10	21–200	35–150	0.10–0.3	20–70

Source: Bergmann, 1983 and Bergmann and Neubert, 1976.

PLANT MICRONUTRIENT REQUIREMENTS

Plant species vary considerably in their requirement for (Römheld and Marschner, 1991) and sensitivity to the micronutrients, as shown in Table 4.7.

In addition, some crop plants are uniquely sensitive to either a deficiency or an excess of a micronutrient; some of these crop plants are listed in Table 4.8.

Another measure of crop sensitivity to a micronutrient can be categorized by the frequency of treatment applied. A listing compiled by Shorrocks (1987), who listed crops by their susceptibility to a particular micronutrient and identified if a micronutrient treatment is normally applied, is given in Table 4.9.

TABLE 4.7 Relative sensitivities of selected crops to micronutrient deficiencies

	Sensitivity to Micronutrient Deficiency					
Crop	*Boron*	*Copper*	*Iron*	*Manganese*	*Molybdenum*	*Zinc*
Alfalfa	High	High	Medium	Medium	Medium	Low
Asparagus	Low	Low	Medium	Low	Low	Low
Barley	Low	Medium	High	Medium	Low	Medium
Bean	Low	Low	High	High	Medium	High
Blueberry	Low	Medium	—[a]	Low	—	—
Broccoli	Medium	Medium	High	Medium	High	—
Cabbage	Medium	Medium	Medium	Medium	Medium	—
Carrot	Medium	Medium	—	Medium	Low	Low
Cauliflower	High	Medium	High	Medium	High	—
Celery	High	Medium	—	Medium	Low	—
Clover	Medium	Medium	—	Medium	Medium	Low
Corn	Low	Medium	Medium	Medium	Low	High
Cucumber	Low	Medium	—	Medium	—	—
Grass	Low	Low	High	Medium	Low	Low
Lettuce	Medium	High	—	High	High	Medium
Oat	Low	High	Medium	High	Low	Low
Onion	Low	High	—	High	High	High
Parsnip	Medium	Medium	—	Medium	—	—
Pea	Low	Low	—	High	Medium	Low
Peppermint	Low	Low	Low	Medium	Low	Low
Potato	Low	Low	—	High	Low	Medium
Radish	Medium	Medium	—	High	Medium	—
Rye	Low	Low	—	Low	Low	Low
Sorghum	Low	Medium	High	High	Low	High
Soybean	Low	Low	High	High	Medium	Medium
Spearmint	Low	Low	—	Medium	Low	Low
Spinach	Medium	High	High	High	High	—
Sudan grass	Low	High	High	High	Low	Medium
Sugar beet	High	Medium	High	Medium	Medium	Medium
Sweet corn	Medium	Medium	Medium	Medium	Low	High
Table beet	High	High	High	High	High	Medium
Tomato	Medium	Medium	High	Medium	Medium	Medium
Turnip	High	Medium	—	Medium	Medium	—
Wheat	Low	High	Low	High	Low	Low

[a] Inadequate available data to categorize into low-, medium-, or high-sensitivity groups.

TABLE 4.8 Crop species sensitive to either deficient or excessive levels of the micronutrients

Micronutrient	Sensitive to Deficiency	Sensitive to Excess
Boron (B)	Legumes, *Brassica* (cabbage and relatives), beets, celery, grapes, fruit trees (apple and pear), cotton, and sugar beet	Cereals, potato, tomato, cucumber, sunflower, and mustard
Chlorine (Cl)	Cereals, celery, potato, coconut palm, sugar beet, lettuce, carrot, and cabbage	Strawberry, navy bean, fruit trees, pea, and onion
Copper (Cu)	Cereals (oat), sunflower, spinach, alfalfa, onion, and watermelon	Cereals and legumes, spinach, citrus seedlings, and gladiolus
Iron (Fe)	Fruit trees (citrus), grapes, several calcifuge species, pecan, sorghum, soybean, and clover	Rice and tobacco
Manganese (Mn)	Cereals (oat), legumes, fruit trees (apple, cherry, and citrus), soybean, and sugar beet	Cereals, legumes, potato, and cabbage
Molybdenum (Mo)	*Brassica* (cabbage and relatives) and legumes	Cereals, pea, and green bean
Zinc (Zn)	Cereals (corn), legumes, grasses, hops, flax, grapes, fruit trees (citrus), soybean, field bean, and pecan	Cereals and spinach

Another measure of the frequency of micronutrient deficiencies worldwide was determined by Sillanpää (1982) in a study of 190 field trials conducted in 115 countries with four major grain crops: maize, rice, wheat, and barley. Data on the estimated degree of deficiency and occurrence are given in Table 4.10.

SOIL AND PLANT SPECIES ASSOCIATIONS

The micronutrients are unique among the essential elements since their deficiency is frequently associated with a combination of crop species and soil characteristics. A partial list of some of the more common crop–soil associations is given in Table 4.11.

TABLE 4.9 Crops susceptible to micronutrient deficiencies (crops which are often treated are underlined)

Boron (B)	Copper (Cu)	Iron (Fe)	Manganese (Mn)	Molybdenum (Mo)	Zinc (Zn)
Alfalfa	Alfalfa	Citrus	Apple	Alfalfa	Apple
Apple	Barley	Field beans	Barley	Beans	Beans
Broccoli	Carrot	Flax	Beans	Broccoli	Citrus
Carrot	Citrus	Grapes	Citrus	Cauliflower	Coffee
Cauliflower	Lettuce	Groundnuts	Grapes	Clovers	Field bean
Celery	Oats	Mint	Lettuce	Lettuce	Flax
Coffee	Onions	Ornamentals	Oats	Peas	Maize
Cotton	Red beet	Sorghum	Peaches	Soybean	Peach
Eucalyptus	Rice	Soybeans	Peas	Spinach	Pear
Grapes	Spinach	Sudangrass	Potato		Pecan
Groundnuts	Tobacco	Fruit trees	Radish		Rice
Oil palm	Wheat	Vegetables	Sorghum		Sorghum
Oilseed rape			Soybean		Soybean
Olive			Spinach		Sudangrass
Pines			Strawberries		Tung
Red + sugar beets			Sudan grass		
Rutabaga			Sugar beet		
Sunflower			Wheat		
Swede and turnip					

Source: Shorrocks, 1987.

TABLE 4.10 Estimated occurrence of micronutrient deficiencies in 15 countries in 190 field trials with maize, rice, wheat, and barley

Micronutrient	Estimated Deficiency Degree and Occurrence (%)		
	Acute	Latent[a]	Total
Boron (B)	10	20	30
Copper (Co)	4	10	14
Iron (Fe)	—	3	3
Manganese (Mn)	1	9	10
Molybdenum (Mo)	3	12	15
Zinc (Zn)	25	24	49

[a] Test crop did not necessarily respond to application of the micronutrient.

Source: Sillanpää, 1982.

TABLE 4.11   Soil conditions and crops where micronutrient deficiencies most often occur

Micronutrient	Sensitive Crops	Soil Conditions for Deficiency
Boron (B)	Alfalfa, clover, cotton, peanut, sugar beet, cabbage	Acid sandy soils low in organic matter, overlimed soils, organic soils
Copper (Cu)	Corn, onions, small grains, watermelon	Organic soils, mineral soil high in pH and organic matter
Iron (Fe)	Citrus, clover, pecan, sorghum, soybean	Leached sandy soils low in organic matter, alkaline soils, soils high in phosphorus
Manganese (Mn)	Alfalfa, small grains, soybean, sugar beet	Leached acid soils, neutral to alkaline soil high in organic matter
Zinc (Zn)	Corn, field beans, pecan, sorghum	Leached acid sandy soils low in organic matter, neutral to alkaline soils and/or high in phosphorus

PHYSIOLOGICAL DISORDERS

Commonly occurring and identified micronutrient deficiencies have been classed as physiological disorders because the visual symptoms are unique in their appearance or the deficiency leads to either biochemical degradation of some plant part or disease infestation, such as is the case for take-all in wheat. Some of the more commonly occurring physiological disorders are given in Table 4.12.

BORON (B)

Function in Plant

Boron is believed to be important in the synthesis of one of the bases for RNA (uracil) formation and in cellular activities (division, differentiation, maturation, respiration, growth, etc.). Boron has long been associated with pollen germination and growth and improves the stability of pollen tubes. Relatively immobile in plants, boron is transported primarily in the xylem.

TABLE 4.12 Physiological disorders that are micronutrient deficiencies

Disorder	Crop	Symptoms	Deficient Micronutrient
Raan (brown heart)	Swede, turnip	Rotting of center of root	Boron
Heart rot	Beets	Death of center of crown, rotting of center of root	Boron
Hollow stem	Cauliflower	Rotting of center of stem	Boron
Bitter pit	Apple	Decay or corking of the flesh under the skin	Boron
Take-all	Wheat	Root rot	Chloride
Wither tip	Cereals	Chlorosis of leaves, withering of tips of leaves and inflorescences	Copper
Whiptail	Cauliflower	Reduction or suppression of leaf blades	Molybdenum
Grey speck	Oat	Irregular grey-brown streaks or specks on leaves	Manganese
Speckled yellows	Sugar beet	Chlorosis between leaf veins, inward curling of leaves	Manganese
Marsh spot	Pea	Brown area in center of seed	Manganese
Little leaf	Apple	Small, malformed leaves, shortened internodes	Zinc

Content and Distribution in Plants

Based on the plant species, boron requirements can be separated into three groups:

- Leaf content of monocots: 1–6 mg/kg boron
- Dicots: 20–70 mg/kg boron
- Dicots with latex system: 80–100 mg/kg boron

Crop sensitivity to boron varies with plant species, as shown in Table

TABLE 4.13   Relative tolerance of plants to boron

Sensitive	American elm, apple, apricot, avocado, blackberry, cherry, cowpea, elm, fig, grape, grapefruit, Jerusalem artichoke, kidney bean, kola, larkspur, lemon, lupine, navy bean, orange, pansy, peach, pear, pecan, persimmon, plum, strawberry, violet, walnut
Semi-tolerant	Alfalfa, barley, bird's-foot trefoil, broccoli, calendula, California poppy, carrot, cauliflower, celery, clover, corn, field pea, hops, Kentucky bluegrass, lettuce, lima bean, millet, milo, mustard, oats, olive, onion, parsley, parsnip, peanut, pepper, potato, pumpkin, radish, rice, rose, rutabaga, spinach, sunflower, sweet corn, sweet pea, sweet potato, timothy, tobacco, tomato, vetch, wheat, zinnia
Tolerant	Artichoke, asparagus, athel, blueberry, chard, cotton, cucumber, gladiolus, mangel, muskmelon, oxalis, palm, pasture grass, peppermint, rye, sesame, soybean, spearmint, Sudan grass, sugar beet, sweet clover, table beet, turnip

4.13. The boron sufficiency range for a number of crop species is given in Table 4.14.

Boron tends to accumulate in leaf margins at concentrations five to ten times that found in the leaf blade. The level can become sufficiently high to result in marginal burning and death of the leaf margin. Boron can exist in the plant as the borate (BO_3^{3-}) anion.

A high calcium content in the plant creates a high boron requirement, while a high potassium plant content accentuates the negative effect of low boron tissue levels.

Available Forms for Root Absorption

Most of the boron in soil exists in the soil's organic plant and microorganism residues; the release of boron by residue decomposition is the major supply source of boron for crop utilization. Boron exists in the soil solution as the borate (BO_3^{3-}) anion. Since the major soil form is undissociated and is neutral in charge, the primary loss of boron from soils is by leaching. Leaching is also a common technique for removing excess boron from the surface soil and rooting zone.

Generally, total boron in the soil can range from 20 to 200 mg/kg, while the amount available for plant absorption ranges from 1 to 5 mg/kg in soil

TABLE 4.14 Sufficiency range for boron in selected crops

Crop	Plant Part	Sampling Time	Sufficiency Range (mg/kg)
Field Crops			
Alfalfa	Top 6 inches	First flower	30–80
Corn	Ear leaf	Initial silk	5–25
Peanut	Upper part of plant	Early pegging	20–50
Rice	Most recent leaf	Panicle initiation	6–7
Soybean	Most recent leaf	Prior to pod set	20–55
Vegetables			
Bean, snap	Upper developed leaves	Initial pod set	20–75
Cucumber, field	Fifth leaf from top	Flower to small fruit set	25–60
Lettuce, cos type	Wrapper leaf	Mature	25–60
Potato, Irish	Upper developed leaf	30-cm tall	40–70
Tomato, field	Adjacent top inflorescence	Mid-bloom	25–60
Watermelon	Fifth leaf from top	Flower start–small fruit	25–60
Fruits and Nuts			
Apple	Mid-shoot current growth	Mid-season	25-50
Banana	6–9-month-old leaves		10-50
Grape	Petiole opposite basal flower duster	Full bloom	25-50
Olive	Mid-shoot		20-75
Orange	Behind fruit	5–7-month leaves	25–100
Peach	Fruiting or non-fruiting spurs	Mid-summer	20-60
Pecan	Mid-portion terminal growth	56–84 days after terminal growth	15–50

solution. The accepted range of boron is narrow, with a deficiency occurring when a hot-water extract contains less than 1 mg/kg. Toxicity occurs at levels above 5 mg/kg.

Boron availability is also influenced by soil water pH; the optimum range for maximum availability lies between 5.5 and 7.0.

Movement in Soil and Root Absorption

Boron moves in soil by mass flow and diffusion, with boron deficiency occurring when soil moisture levels are low for extended periods or after a long period of heavy leaching rainfall.

Deficiency Symptoms

Plants deficient in boron exhibit an abnormal growth of growing points (meristematic tissue), with apical growing points eventually becoming stunted, which may then eventually die. When the plant is boron deficient, auxins accumulate at the growing points, and leaves and stems will become brittle.

Excess (Toxicity) Symptoms

An excess of boron causes leaf tips to become yellow, followed by necrosis. Leaves eventually assume a scorched appearance and prematurely fall off.

Fertilizer Sources

Boron can be either soil or foliar applied; rates of application range from 0.56 to 2.2 kg/ha (0.5 to 2.0 lb/acre) for soil application. Care is needed when applying boron since toxicity can occur from irregular soil distribution. A list of boron fertilizer sources is given in Table 4.15.

CHLORINE (Cl)

Function in Plants

Chlorine is involved in the evolution of oxygen (O_2) in photosystem II in the process of photosynthesis. Chlorine raises the cell osmotic pressure,

TABLE 4.15   Micronutrient-containing materials, their formula, and micronutrient content

Micronutrient Sources	Formula	Micronutrient Content (%)
Boron (B)		
Fertilizer borate—48	$Na_2B_4O_7 \cdot 10H_2O$	14–15
Fertilizer borate—granular	$Na_2B_4O_7 \cdot 10H_2O$	14
Foliarel	$Na_2B_8O_{13} \cdot 4H_2O$	21
Solubor	$Na_2B_4O_7 \cdot 4H_2O + Na_2B_{10}O_{16} \cdot 10H_2O$	20.5
Borax	$Na_2B_4O_7 \cdot 10H_2O$	11
Copper (Cu)		
Copper sulfate (monohydrate)	$CuSO_4 \cdot H_2O$	35
Copper sulfate (pentahydrate)	$CuSO_4 \cdot 5H_2O$	25
Cupric oxide	CuO	75
Cuprous oxide	Cu_2O	89
Cupric ammonium phosphate	$Cu(NH_4)PO_4 \cdot H_2O$	32
Basic copper sulfates	$CuSO_4 \cdot 3Cu(OH)_2$ (general formula)	13–53
Cupric chloride	$CuCl_2$	17
Copper chelates	$Na_2CuEDTA$	13
	$NaCuHEDTA$	9
Copper polyflavonoid	Organically bound Cu	5–7
Iron (Fe)		
Ferrous ammonium phosphate	$Fe(NH_4)PO_4 \cdot H_2O$	29
Ferrous ammonium sulfate	$(NH_4)_2SO_4 \cdot FeSO_4 \cdot 6H_2O$	14
Ferrous sulfate	$FeSO_4 \cdot 7H_2O$	19–21
Ferric sulfate	$Fe_2(SO_4)_3 \cdot 4H_2O$	23
Iron chelates	$NaFeEDTA$	5–11
	$NaFeHFDTA$	5–9
	$NaFeEDDHA$	6
	$NaFeDTPA$	10
Iron polyflavonoids	Organically bound Fe	9–10
Manganese (Mn)		
Manganese sulfate	$MnSO_4 \cdot 4H_2O$	26–28
Manganese oxide	MnO	41–68
Manganese chelate	Mn-EDTA	5–12
Molybdenum (Mo)		
Ammonium molybdate	$(NH_4)_6Mo_7O_{24} \cdot 2H_2O$	54
Sodium molybdate	$Na_2MoO_4 \cdot 2H_2O$	39–41
Molybdenum trioxide	MoO_3	66

TABLE 4.15 **Micronutrient-containing materials, their formula, and micronutrient content (continued)**

Micronutrient Sources	Formula	Micronutrient Content (%)
Zinc (Zn)		
Zinc sulfate	$ZnSO_4 \cdot 7H_2O$	35
Zinc oxide	ZnO	78–80
Zinc chelates	$Na_2ZnEDTA$	14
	$NaZnNTA$	13
	$NaZnHEDTA$	9
Zinc polyflavonoids	Organically bound Zn	10

affects stomatal regulation, and increases the hydration of plant tissue. Chlorine may be related to the suppression of leaf spot and take-all disease in wheat.

Content and Distribution in Plants

The leaf content of chlorine ranges from low parts per million levels (20 mg/kg) in dry matter to percent concentrations. A deficiency occurs in wheat when plant levels are less than 0.15%. Chlorine exists in plants as the chloride (Cl^-) anion.

Available Forms for Root Absorption

Chlorine exists in soil solution as the chloride (Cl^-) anion, which moves in the soil by mass flow. The Cl^- anion competes with other anions, such as nitrate (NO_3^-) and sulfate (SO_4^{2-}), for uptake.

Deficiency Symptoms

Typically, plants deficient in chlorine exhibit a chlorosis of younger leaves and wilting of the plant. In wheat, chlorine deficiency has been related to disease infestation called take-all. Chlorine may affect other grain crops as well.

Excess (Toxicity) Symptoms

An excess of chlorine results in a premature yellowing of the leaves, burning of the leaf tips and margins, bronzing, and abscission of the leaves (due to salt effect).

Fertilizer Sources

The use of chloride-containing fertilizers, such a potassium chloride (KCl), plus the trace amounts of chlorine found in many fertilizer materials would be sufficient in most cases to satisfy the crop requirement for chlorine.

COPPER (Cu)

Function in Plants

Copper is a constituent of the chloroplast protein plastocyanin, as well as serving as a part of the electron transport system linking photosystem I and II. This element participates in protein and carbohydrate metabolism and nitrogen (N_2) fixation. It is a part of the enzymes cytochrome oxidase, ascorbic acid oxidase, and polyphenol oxidase, which reduce both atoms of molecular oxygen (O_2). Also, copper is involved in the desaturation and hydroxylation of fatty acids.

Content and Distribution in Plants

The copper sufficiency range in leaves is between 3 and 7 mg/kg of the dry matter, while the toxicity range begins at 20 to 30 mg/kg. Much higher values, 20 to 200 mg/kg, can be tolerated if copper has been applied as a fungicide.

In general, copper deficiency is not likely to occur since the copper requirement for most crops is quite low, unless the conditions given in Table 4.11 exist. The relative tolerance of crop plants to copper is given in Table 4.16. Sufficiency ranges for a number of crop species are listed in Table 4.17.

Copper in the plant can interfere with iron metabolism, which may result in the development of an iron deficiency. In its interaction with molybdenum, it may interfere with the enzymatic reduction of nitrate (NO_3). Grasses and forages with unusually high and/or low ratios of copper to molybdenum may pose significant health problems for ruminant animals.

TABLE 4.16   Relative tolerance of plants to copper

Low	Asparagus, beans, *Lotus* spp., lupine, pasture grass, pea, peppermint, pine, potato, rape, rice, rutabaga, rye, soybean, spearmint
Moderate	Apple, barley, blueberry, broccoli, cabbage, cauliflower, celery, corn, cotton, cucumber, mangel, parsnip, peach, pear, pineapple, pome and stone fruits, radish, sorghum, strawberry, sugar beet, sweet corn, Swiss chard, tomato, tung oil, turnip, vines
High	Alfalfa, barley, carrot, citrus, dill, lettuce, lucerne, millet, oat, onion, pangolagrass, spinach, Sudan grass, sunflower, table beet, wheat

Available Forms for Root Absorption

Copper exists in the soil primarily in complexed forms as low molecular weight organic compounds, such as humic and fulvic acids. Cupric ion (Cu^{2+}) is present in very small quantities in soil solution, with copper deficiency occurring primarily on sandy and organic soils. Copper uptake rates are lower than for most other micronutrients.

Movement in Soil and Root Absorption

Although the copper supply in soil solution is very low, most soils have sufficient copper to meet crop requirements. Most soils are able to maintain sufficient Cu^{2+} ions in soil solution even with increasing soil pH.

Deficiency Symptoms

Symptoms of copper deficiency are reduced or stunted growth with distortion of young leaves and necrosis of the apical meristem. In trees, copper deficiency may cause white tip or bleaching of younger leaves and summer dieback.

Excess (Toxicity) Symptoms

An excess of copper can induce iron deficiency and chlorosis. Root growth may be suppressed, with inhibited elongation and lateral root formation.

TABLE 4.17　Sufficiency range for copper in selected crops

Crop	Plant Part	Sampling Time	Sufficiency Range (mg/kg)
Field Crops			
Alfalfa	Top 6 inches	First flower	7–30
Corn	Ear leaf	Initial silk	6–20
Peanut	Upper part of plant	Early pegging	10–50
Rice	Most recent leaf	Panicle initiation	8–25
Soybean	Most recent leaf	Prior to pod set	10–30
Winter wheat	Top two leaves	Just before heading	5–50
Vegetables			
Bean, snap	Upper developed leaves	Initial pod set	7–30
Cucumber, field	Fifth leaf from top	Flower to small fruit set	7–20
Lettuce, cos type	Wrapper leaf	Mature	5–25
Potato, Irish	Upper developed leaf	30 cm tall	7–20
Tomato, field	Adjacent top inflorescence	Mid-bloom	5–20
Watermelon	Fifth leaf from top	Flower start–small fruit	6–20
Fruits and Nuts			
Apple	Mid-shoot current growth	Mid-season	6–50
Banana	6–9-month-old leaves		6–25
Orange	Behind fruit	5–7-month leaves	6–100
Peach	Fruiting or non-fruiting spurs	Mid-summer	5–16
Pecan	Mid-portion terminal growth	56–84 days after terminal growth	6–30

Fertilizer Sources

Copper can be either soil or foliar applied, using the sources given in Table 4.15.

IRON (Fe)

Function in Plant

Iron is an important component in many plant enzyme systems, such as cytochrome oxidase (electron transport) and cytochrome (terminal respiration step). Iron is a component of protein ferredoxin and is required for nitrate (NO_3) and sulfate (SO_4) reduction, nitrogen (N_2) assimilation, and energy (NADP) production. It functions as a catalyst or part of an enzyme system associated with chlorophyll formation. It is thought that iron is involved in protein synthesis and root-tip meristem growth.

Content and Distribution

Leaf iron content ranges from 10 to 1000 mg/kg in dry matter, with sufficiency ranging from 50 to 75 mg/kg, although total iron may not always be related to sufficiency. In general, the 50-mg/kg iron level is the generally accepted critical value for most crops.

Iron deficiency affects many crops. A common deficiency that occurs on alkaline soils is lime chlorosis. A list of ornamental plants intolerant to high-lime soils is given in Table 4.18.

The majority of plant iron is in the ferric (Fe^{3+}) form as ferric phosphoprotein, although the ferrous (Fe^{2+}) ion is believed to be the metabolically active form.

High phosphorus decreases the solubility of iron in the plant; a P:Fe ratio of 29:1 is average for most plants. Potassium increases the mobility and solubility of iron, while nitrogen accentuates iron deficiency due to increased growth. The bicarbonate (HCO_3^-) anion is believed to interfere with iron translocation. High zinc can interfere with iron metabolism, resulting in a visual symptom of iron deficiency.

Extractable ferrous (Fe^{2+}) iron may be a better indicator of plant iron status than total iron. Various extraction procedures have been proposed for diagnosing iron deficiency, with 20 to 25 mg/kg as the critical extractable iron range. However, most methods of determining the iron status of plants

TABLE 4.18   Ornamental plants intolerant to high-lime soils

Ground covers	Dichondra, strawberry, spring cinquefoil, African daisy, star daisy, periwinkle, and many turf species
Shrubs and trees	Abelia, acacia, azalea, bottlebrush, camellia, catalpa, citrus, eucalyptus, hibiscus, hydrangea, black walnut, sweet gum, crabapple, photina, Monterey pine, evergreen pear, firethorn, rhododendron, rose, weeping willow, western red cedar, viburnum, wisteria, peach, gardenia, raspberry, juniper, spirea, privet, lilac, honeysuckle, teatree, avocado, magnolia, pin oak, heavenly bamboo
Other herbaceous plants	Petunia, peony, iris, gladiolus, geranium, lupine, verbena, ferns

by means of an iron analysis, whether total or extractable, are flawed. The indirect measurement of chlorophyll content may be the best alternative method for diagnosing iron sufficiency.

Available Forms for Root Absorption

Iron exists in the soil as both ferric (Fe^{3+}) and ferrous (Fe^{2+}) cations. The Fe^{2+} form, whose availability is affected by the degree of soil aeration, is thought to be the active form taken up by plants. Iron-sufficient plants can acidify the rhizosphere as well as release iron-complexing substances, such as siderophores, which enhance availability and uptake. Iron-inefficient plants do not have roots that will readily acidify the rhizosphere and/or release iron-complexing substances. A list of iron-sensitive crops is given in Table 4.19.

Movement in Soil and Root Absorption

Iron moves in the soil by mass flow and diffusion, primarily as the ferric (Fe^{3+}) cation, which, when it enters the root rhizosphere, can be reduced to the ferrous (Fe^{2+}) form, released from any chelate form that exists, and is then root absorbed. Copper, manganese, and calcium competitively inhibit iron uptake, and high levels of phosphorus will also reduce iron uptake.

TABLE 4.19   Iron-sensitive crops

Fruit and nut crops	Apple, apricot, avocado, banana, blueberry, brambles, cacao, cherry, citrus, coconut, coffee, grape, nuts (almond, filbert, pecan, walnut), olive, peach, pear, pineapple, plum, strawberry, tung
Others	Corn, rice, sorghum, soybean, sugarcane

Deficiency Symptoms

Interveinal chlorosis of younger leaves is the typical symptom of deficiency. Appearance may be similar to manganese deficiency, which makes verification essential. As the severity of the deficiency increases, chlorosis spreads to the older leaves, and the entire plant may become chlorotic.

Excess (Toxicity) Symptoms

Iron may accumulate to several hundred milligrams per kilogram without symptoms of toxicity. Toxicity produces a bronzing of the leaves with tiny brown spots, which frequently occurs in rice.

Fertilizer Sources

Iron sources can be either soil or foliar applied. Foliar application is the most efficient, using either a solution of ferrous sulfate ($FeSO_4$) or one of the chelated (EDTA or EDDHA) forms of iron. A list of iron sources is given in Table 4.15.

MANGANESE (Mn)

Function in Plants

Manganese is involved in the oxidation-reduction processes in the photosynthetic electron transport system. It is essential in the photosystem II for photolysis, acts as a bridge for ATP and enzyme complex phosphokinase and phosphotransferases, and activates IAA oxidases.

Content and Distribution

The leaf sufficiency content of manganese ranges from 10 to 50 mg/kg in dry matter in mature leaves. Tissue levels will reach 200 mg/kg or higher (soybean, 600 mg/kg; cotton, 700 mg/kg; sweet potato, 1380 mg/kg) before severe toxicity symptoms develop. Relative sensitivity to manganese by crop species is given in Table 4.20, and sufficiency range by crop species is given in Table 4.21.

TABLE 4.20   Relative sensitivity of plants to manganese

Low sensitivity	Asparagus, blueberry, cotton, rye
Moderate sensitivity	Alfalfa, barley, broccoli, cabbage, carrot, cauliflower, celery, clover, corn, cucumber, grass, parsnip, peppermint, sorghum, spearmint, sugar beet, tomato, turnip
High sensitivity	Beans, citrus, lettuce, oats, onion, pea, peach, potato, radish, soybean, spinach, Sudan grass, table beet, wheat

Available Forms for Roots Absorption

Manganese exists in the soil solution as either Mn^{2+}, Mn^{3+}, or Mn^{4+} cations and as exchangeable manganese. The cation Mn^{2+} is the ionic form taken up by plants. Availability is significantly affected by soil pH; it decreases when the pH increases above 6.2 in some soils, while in other soils the decrease may not occur until the soil water pH reaches 7.5. Manganese toxicity is frequently associated with low (<5.5) soil pH. Manganese availability can be reduced significantly by low (<70°F) soil temperatures.

Movement in Soil and Root Absorption

Manganese is primarily supplied to the plant by mass flow and root interception. Low soil temperature and moisture stress will reduce manganese uptake. Some plants may release root exudates that reduce Mn^{4+} to Mn^{2+}, complex it, and then make it more available for root uptake. Manganese is not known to interfere with the metabolism or uptake of any of the other essential elements.

TABLE 4.21 Sufficiency range for manganese in selected crops

Crop	Plant Part	Sampling Time	Sufficiency Range (mg/kg)
Field Crops			
Alfalfa	Top 6 inches	First flower	30–100
Corn	Ear leaf	Initial silk	20–200
Peanut	Upper part of plant	Early pegging	100–350
Rice	Most recent leaf	Panicle initiation	150–800
Soybean	Most recent leaf	Prior to pod set	20–100
Winter wheat	Top two leaves	Just before heading	16–200
Vegetables			
Bean, snap	Upper developed leaves	Initial pod set	50–300
Cucumber, field	Fifth leaf from top	Flower to small fruit set	50–300
Lettuce, cos type	Wrapper leaf	Mature	11–250
Potato, Irish	Upper developed leaf	30-cm tall	30–250
Tomato, field	Adjacent top inflorescence	Mid-bloom	40–250
Watermelon	Fifth leaf from top	Flower start–small fruit	50–250
Fruits and nuts			
Apple	Mid-shoot current growth	Mid-season	25–200
Banana	6–9-month-old leaves		100–1000
Grape	Petiole opposite basal flower cluster	Full bloom	18–100
Olive	Mid-shoot		25+
Orange	Behind fruit	5–7-month leaves	25–200
Peach	Fruiting or non-fruiting spurs	Mid-summer	40–160
Pecan	Mid-portion terminal growth	56–84 days after terminal growth	200–500

Deficiency Symptoms

Reduced or stunted growth with dicots showing interveinal chlorosis of the younger leaves is symptomatic of manganese deficiency. In some cases, manganese deficiency may be confused with iron deficiency symptoms, requiring verification by means of plant analysis. Verification of manganese deficiency can be done by spotting the chlorotic leaf or dipping the leaf into a manganese-containing solution. Chlorosis under the treated area or of the entire leaf will disappear in about five days if manganese is the deficient micronutrient. Cereals develop gray spots on their lower leaves (grey speck), and legumes develop necrotic areas on their cotyledons (marsh spot).

Excess (Toxicity) Symptoms

An excess of manganese causes older leaves to show a brown spot surrounded by a chlorotic zone or circle. Black specks on stone fruits, particularly in apple, and referred to as measles, are the result of high manganese content in the tissue.

Fertilizer Sources

Manganese is best applied as a foliar spray to correct a deficiency, as soil applications can be very inefficient due to soil inactivation of applied manganese. Row application of a phosphorus fertilizer will increase manganese availability and uptake, which may be sufficient to avoid the occurrence of a manganese deficiency. A list of manganese sources is given in Table 4.15.

MOLYBDENUM (Mo)

Function in Plants

Molybdenum is a component of two major enzyme systems, nitrogenase and nitrate reductase. Nitrogenase is involved in the conversion of the nitrate (NO_3^-) anion to the ammonium (NH_4^+) cation. The requirement for molybdenum is reduced greatly by the availability and utilization of NH_4.

Content and Distribution

The leaf content of molybdenum is usually less than 1 mg/kg in dry matter, due in part to the very low level of the molybdate (MnO_4^{2-}) anion in soil solution.

Molybdenum can be taken up in higher amounts without resulting in toxic effects to the plant. However, high molybdenum content (>10 mg/kg) forage can pose a health hazard to cattle. The normal molybdenum plant content ranges from 0.34 to 1.5 mg/kg.

Available Forms for Root Absorption

The primary soluble soil form is the molybdate (MoO_4^{2-}) anion, whose availability is increased tenfold for each unit increase in soil pH. Molybdenum is strongly absorbed by iron and aluminum oxides in the soil.

Movement in Soil and Root Absorption

Mass flow and diffusion each supply molybdenum to roots, with mass flow supplying most of the molybdenum when soil molybdenum level is high.

If nitrate (NO_3) is the primary nitrogen source, molybdenum uptake is higher. In general, phosphorus and magnesium will enhance molybdenum uptake, whereas sulfate (SO_4) will reduce uptake.

Deficiency Symptoms

Molybdenum deficiency symptoms frequently resemble nitrogen deficiency symptoms. Older and middle leaves become chlorotic first, and in some instances leaf margins are rolled and growth and flower formation are restricted. Cruciferae and pulse crops have high molybdenum requirements (see Tables 4.7 to 4.9). In cauliflower, the middle lamella of the cell wall is not formed completely, and only the leaf rib is formed, giving a whip tail appearance in severe cases.

Excess (Toxicity) Symptoms

High plant molybdenum does not normally affect the plant but can pose a problem for ruminant animals that consume plants containing 5 mg/kg or more molybdenum.

Fertilizer Sources

Molybdenum is best supplied by means of seed treatment. The sources of molybdenum are given in Table 4.15.

ZINC (Zn)

Function in Plant

Zinc is involved in the same enzymatic functions as are manganese and magnesium. Only carbonic anhydrase has been found to be specifically activated by zinc.

Content and Distribution

The leaf sufficiency content of zinc ranges from 15 to 50 mg/kg in dry matter in mature leaves, while with some species deficiency will not occur until the zinc content is as low as 12 mg/kg. For many crops, 15 mg/kg in the leaves is considered the critical zinc level. A small variation in zinc content, as little as 1 to 2 mg/kg at the critical level, may be sufficient to distinguish between deficiency and sufficiency. Some plants can accumulate considerable quantities of zinc (several hundred milligrams per kilogram) without harm to the plant, although high zinc can induce iron deficiency in some crops. High soil zinc content, the result of sewage sludge application, will significantly reduce seed generation for many crops. A list of the sufficiency ranges for a number of crops is given in Table 4.22.

The relationship between phosphorus and zinc has been intensively studied, as research suggests that high phosphorus can interfere with zinc metabolism as well as with zinc uptake through the root. High zinc will induce an iron deficiency in many plants, particularly those sensitive to iron (see Table 4.19).

Available Forms for Root Absorption

Zinc exists in soil solution as the Zn^{2+} cation, as exchangeable zinc, and as organically complexed zinc. Availability is affected by soil pH, decreasing with increasing pH. Zinc availability can also be reduced when the available soil phosphorus level is very high.

TABLE 4.22 Sufficiency range for zinc in selected crops

Crop	Plant Part	Sampling Time	Sufficiency Range (mg/kg)
Field Crops			
Alfalfa	Top 6 inches	First flower	20–70
Corn	Ear leaf	Initial silk	25–100
Peanut	Upper part of plant	Early pegging	20–50
Rice	Most recent leaf	Panicle initiation	18–50
Soybean	Most recent leaf	Prior to pod set	20–50
Winter wheat	Top two leaves	Just before heading	20–70
Vegetables			
Bean, snap	Upper developed leaves	Initial pod set	20–200
Cucumber, field	Fifth leaf from top	Flower to small fruit set	25–100
Lettuce, cos type	Wrapper leaf	Mature	20–250
Potato, Irish	Upper developed leaf	30-cm tall	30–200
Tomato, field	Adjacent top inflorescence	Mid-bloom	20–50
Watermelon	Fifth leaf from top	Flower start–small fruit	20–50
Fruits and nuts			
Apple	Mid-shoot current growth	Mid-season	20–100
Banana	6–9-month-old leaves		20–200
Grape	Petiole opposite basal flower cluster		20–30
Olive	Mid-shoot	Full bloom	25+
Orange	Behind fruit	5–7-month leaves	25–200
Peach	Fruiting or non-fruiting spurs	Mid-summer	20–50
Pecan	Mid-portion terminal growth	56–84 days after terminal growth	50–100

Movement in Soil and Root Absorption

Zinc is brought into contact with plant roots by mass flow and diffusion, with diffusion being the primary delivery method. Copper (Cu^{2+}) and the other cations, such as the ammonium (NH_4^+) cation, will inhibit root zinc uptake. Phosphorus appears to inhibit translocation rather than directly inhibiting uptake.

The efficiency of zinc uptake seems to be enhanced by a reduction of the pH in the rhizosphere. Those plant species that effectively reduce the pH are less affected by high soil zinc than those plant species that do not.

Deficiency Symptoms

Zinc deficiency appears as a chlorosis in the interveinal areas of new leaves, producing a banding appearance. With increasing severity of the deficiency, leaf and plant growth become stunted (rosette), and leaves die and fall off the plant. At branch terminals of fruit and nut trees, rosetting occurs, with considerable dieback at the ends of the branches.

Excess (Toxicity) Symptoms

Plants particularly sensitive to iron (see Table 4.19) will become chlorotic when zinc levels are abnormally high (>400 mg/kg).

Fertilizer Sources

Zinc can be applied to the plant by means of both soil and foliar applications. A list of zinc sources is given in Table 4.15.

5

SOIL TESTING FOR THE MICRONUTRIENTS

INTRODUCTION

Soil testing for the micronutrients has had an interesting history of development and use. Many of the testing procedures were developed for very specific purposes but over the years have been adapted for wider applications. For example, the hot-water soil test for boron was initially developed by Wear (1965) to predict the need for boron on sandy acid soils in Alabama for the production of cotton. Since its initial introduction, the test method has been modified in terms of the laboratory technique for making the assay as well as the application of the results to soils and crops other than initially specified. Therefore, when using any soil test method for a micronutrient or group of micronutrients, the user needs to be aware of either the conditions under which the test was initially designed or conditions for its current application (Cox and Kamprath, 1972; De Temmerman et al., 1984; McGrath et al., 1985; Whitney, 1988; Johnson and Fixen, 1990; Martens and Lindsay, 1990; Sims, 1991; Sims and Johnson, 1991; Gupta, 1993; Liang and Karamanos, 1993).

A micronutrient soil test has two important aspects:

- The laboratory test procedure
- The interpretation of the test result

The test result is no better than its ability to correctly identify the micronutrient status of a soil in terms of probable crop response. Although a micronutrient soil test may be well correlated to crop response, the user is frequently only interested in knowing whether the soil is either sufficient or not sufficient in order to determine if supplementation will be required. A more recent concern is that a micronutrient soil test level may be at a toxic concentration, an aspect that is not always known in most tests.

The soil chemistry associated with most micronutrients is very complex. Most of the micronutrients exist in the soil in many different forms, such as a constituent of the soil organic matter, and therefore only released upon organic matter decomposition, as well as being chelated by soil humus. In addition, availability for root uptake is affected by the pH of the soil, the degree of soil aeration, the soil moisture level, and soil temperature. Some micronutrient soil tests factor in some of these correlated soil conditions when evaluating the micronutrient status of a soil.

MICRONUTRIENT SOIL TESTS

In general, most micronutrient soil test extractants are acids (i.e., 0.1 M HCl, 0.05 M HCl, 0.033 M H_3PO_4), are so-called universal extractants (i.e., Mehlich No. 1 and Mehlich No. 3), or are a chelate in solution used either alone (DTPA) or in some combination with an acid (i.e., DTPA-TEA) or salt solution (ammonium bicarbonate-DTPA, acidic ammonium acetate [AAAc-EDTA], modified Olsen, and Wolf-Morgan). Therefore, there are a number of different micronutrient soil testing procedures which have been designed for specific soil–crop situations (see Table 4.12). Some of the more commonly used soil test procedures for micronutrient determination are given in Table 5.1. Note that there are specific characteristics associated with these testing procedures which are normally applied in the interpretation of the test result obtained.

In a study conducted by Sillanpää (1982), some of the commonly used micronutrient soil tests used in many different countries were categorized, classifying the test results on the basis of five fertility classes, as shown in Table 5.2.

TABLE 5.1 **Soil test methods, soil factors influencing their interpretation, and typical ranges in critical levels for the micronutrients**

Micronutrient	Interacting Factors	Method	Range in Critical Level (mg/kg)
Boron	Crop yield goal, pH, soil moisture, texture, organic matter, soil type	Hot-water soluble	0.1–2.0
Copper	Crop, organic matter, pH, percent CaCO$_3$	Mehlich 1 and Mehlich 3	0.1–10.0
		DTPA	0.1–2.5
		AB-DTPA	>0.5
		0.1 M HCl	1.0–2.0
		Modified Olsen's	0.3–1.0
Iron	pH, percent CaCO$_3$, aeration, soil moisture, organic matter, CEC	DTPA	2.5–5.0
		AB-DTPA	4.0–5.0
		Modified Olsen's	10.0–16.0
Manganese	pH, texture, organic matter, percent CaCO$_3$	Mehlich 1	5.0 at pH 6 / 10.0 at pH 7
		Mehlich 3	4.0 at pH 6 / 8.0 at pH 7
		DTPA	1.0–5.0
		0.1 M HCl	1.0–4.0
		0.03 M H$_3$PO$_4$	10.0–20.0
		Modified Olsen's	2.0–5.0
Molybdenum	pH, crop	Ammonium oxalate, pH 3.3	0.1–0.3
Zinc	pH, percent CaCO$_3$, P, organic matter, percent clay, CEC	Mehlich 1	0.5–3.0
		Mehlich 3	1.0–2.0
		DTPA	0.2–2.0
		AB-DTPA	0.5–1.0
		Modified Olsen's	1.5–3.0
		0.1 M HCl	1.0–5.0

Source: Sims and Johnson, 1991.

Based on data gathered from different worldwide sources, Sillanpää (1982) has determined the range in deficiency and excess for seven micronutrient soil testing procedures (Table 5.3). Note that pH corrections are required for the manganese, molybdenum, and zinc tests. In 1990, Sillanpää (1990) published a summary of the micronutrient assessment at the country level.

TABLE 5.2   Classification of micronutrient contents in five fertility classes

Micronutrient (Test Method)	mg/L				
	Class 1 (Very Low)	Class 2 (Low)	Class 3 (Medium)	Class 4 (High)	Class 5 (Very High)
Boron (hot water)	<0.15	0.15	0.35	0.80	2.0
Copper (AAAc-EDTA)[a]	<0.7	0.7	2.0	6.0	18.0
Iron (AAAc-EDTA)	<30	30	75	200	500
Manganese (Mn)					
(AAAc-EDTA)	<23	23	90	360	1400
(DTPA)	<4	4	14	50	170
Molybdenum (Mo)					
(AAAc-EDTA)	<0.003	0.003	0.14	0.065	0.3
Zinc (Zn)					
(AAAc-EDTA)	<0.5	0.5	1.5	5.0	15.0
(DTPA)	<0.2	0.2	0.7	2.4	8.0

Note: Class 1 and 5 represent the 3 to 4% of the lowest and highest values, respectively, while the remaining middle classes are equally divided with that remaining.

[a] AAAc = acidified ammonium acetate.

Source: Sillanpää, 1982.

TABLE 5.3   Range in deficiency and excess for seven micronutrient soil tests

Micronutrient and Method	Range of (mg/L)	
	Deficiency	Excess
Boron, hot-water extraction	<0.3–0.5	>3–5
Copper, AAAc[a]-EDTA	<0.8–1.0	>17–25
Iron, AAAc-EDTA <30–35	—	
Manganese, DTPA + pH correction	<2–4	>150–200
Manganese, AAAc-EDTA + pH correction	<10–25	1300–2000
Molybdenum, AAAc-EDTA + pH correction	<0.002–0.005	0.3–1.0
Zinc, DTPA	<0.4–0.6	>10–2
AAAc-EDTA + pH correction	<1.0–1.5	>20–30

[a] AAAc = acidified ammonium acetate.

Source: Sillanpää, 1982.

In the IFA World Fertilizer Use Manual (Halliday and Trenkel, 1992), interpretation data for some of the commonly used soil testing procedures for four micronutrients were given (Table 5.4). Note that pH correction is required for the manganese soil test. The interpretative values given in this table are in close agreement with those found in Tables 5.1 and 5.3.

TABLE 5.4   Interpretation of soil data for micronutrients

		Available Nutrients (mg/L of soil = 1.5 kg min. soil)		
		Ranges of Supply		
Micronutrient	*Extraction Method*	*Deficient*	*Medium*	*Excess*
Boron (B)	hot water	<0.3–0.5	(0.5–2)	>3–5
Copper (Cu)	AAAc-EDTA	<0.8–1.0	(2–10)	>17–25
Manganese (Mn)	AAAc + pH correction	<10–25	(100–500)	>1300
Zinc (Zn)	DTPA	<0.4–0.6	(1–5)	>10–20

Note: Hot water = Berger and Truog (1939). AAAc = acid ammonium acetate (0.5 *M* ammonium acetate + acetic acid; pH 4.7), with 0.02 *M* Na-EDTA (Lakanen and Ervio, 1971) (pH correction for Mn: at pH 6.5, factor 0.5; at pH 6, factor 1; at pH 5.5, factor 1.5). DTPA = 0.005 *M*, pH 7.3 (Lindsay and Norwell, 1978).

Source: Halliday and Trenkel, 1992.

TESTING SOILS FOR THE TOXIC METALS

With the increasing use of land application as a disposal method for waste products, particularly sewage sludge, there is a need for soil testing procedures that can be used to determine both the toxic metal level in the soil as well as the potential load limit. Elements of primary interest are cadmium, chromium, mercury, nickel, and lead, although several of the micronutrients, particularly zinc and copper, would also be of interest. The only test procedure that has been designed specifically for such an application is the Baker Test (Baker, 1971, 1973; Baker and Amacher, 1981; Jones, 1992). The test procedure is based on the concept of ion activities in an equilibrium solution which can be used to determine the immediate availability or intensity within the soil–water system. The method requires the determination of the major, micronutrient, and trace element content of the soil and a computer program to interpret the analytical results obtained. The method has been primarily used for assaying disturbed strip-land soils and soils treated with sewage sludge.

In addition, load limits have been defined for the heavy metals cadmium, copper, lead, nickel, and zinc based the cation exchange capacity (CEC) of the soil (Logan and Chaney, 1983) (Table 5.5). Therefore, soil tests for those trace elements considered toxic at elevated concentrations need to be defined in terms of acceptable limits based on the CEC of the soil.

TABLE 5.5   Cumulative metal (trace element) additions in soil based on soil CEC

	Cumulative Metal Additions (kg/ha)		
Trace Element	<5 meq/100 g	5–15 meq/100 g	>15 meq/100 g
Cadmium (Cd)	5.6	11.2	22.4
Nickel (Ni)	56	112	224
Copper (Cu)	140	280	560
Zinc (Zn)	280	560	1120
Lead (Pb)	560	1120	2240

Source: Logan and Chaney, 1983.

FUTURE DEVELOPMENTS

Most of the micronutrient soil tests in use today were developed a number of years ago, and there seems to be little effort being made to develop new tests or to improve current testing procedures. Considerable efforts, however, are being made to utilize trace element analytical data obtained with the use of universal soil extractants, such as Mehlich No. 1 and 3 and the ammonium bicarbonate-DTPA, since these extractants are being assayed by multielement analytical procedures which include the micronutrients and some trace elements. Examples of these efforts can be found in the published work by Jones (1989) using the Mehlich No. 3 extractant for the determination of copper and zinc in Hungarian soils; Shuman et al. (1992) comparing Mehlich No. 1 and 3 extractions with hot-water extractable boron, Boon and Soltanpour (1991) using the ammonium bicarbonate-DTPA extractant to determine cadmium, lead, and zinc in contaminated soils; and the same extractant for determining the selenium status of soils (Soltanpour and Workman, 1980). Such comparisons will continue, as the effort to develop specific micronutrient or trace element tests seems remote at this time.

6

ANALYTICAL PROCEDURES FOR TRACE ELEMENT DETERMINATIONS

PROGRESS IN ANALYTICAL TECHNIQUES

Trace element research has been markedly advanced by the development of sensitive instrumental analytical techniques for the determination of minute trace element levels in a wide range of substances. Prior to the 1970s, much of the limitation associated with trace element research was the inability of ecologists, biologists, and animal and human physiologists to easily assay large numbers of samples at elemental detection limits sufficient to uncover significant effects. Up to that time, most of the analytical procedures for determining trace element contents were tedious and frequently difficult to perform, thereby requiring considerable skill and experience on the part of the analyst. The common analytical techniques were UV-VIS spectrophotometric (colorimetric) methods (Piper, 1942; Rump and Krist, 1988) and both AC (Conner and Bass, 1959) and DC (Thompson and Blanston, 1969) arc emission spectroscopy (Mitchell, 1964).

Today, these same scientists can select from a number of instrumental procedures that will provide reliable analytical information at a reasonable cost and effort. The techniques in common use are atomic absorption (flame and flameless) spectrometry (AAS) (Baker and Suhr, 1982; Tsalev, 1984; Metcalfe, 1987; Ure, 1991), DC plasma emission (DCP) (Skogerboe et al., 1976) and inductively coupled plasma emission spectrometry (ICP-AES) (Barnes, 1978; Montaser and Golightly, 1987; Soltanpour et al., 1982; Thompson and Walsh, 1983; Zarcinas, 1984; Varma, 1991; Sharp, 1991), inductively coupled plasma emission-mass spectrometry (ICP-MS) (Günter et al., 1992; Grosser, 1995), and X-ray fluorescence spectrometry (TXRF) (Kubota and Lazar, 1971; A.A. Jones, 1982, 1991; Prange and Schwenke, 1992; Markert et al., 1994). An additional method is neutron activation analysis (NAA) (Haskin and Ziege, 1971; Helmke, 1982; Arambel et al., 1986; Salmon and Cawse, 1991), which is a very sensitive analytical technique but requires unique specialized equipment. Anodic stripping voltammetry (Huiliang et al., 1987) and differential pulse polarography (Street and Peterson, 1982) are two additional analytical procedures that have unique applications for specific determinations.

The method of analysis chosen depends on many factors which have been discussed by Morrison (1979), Hislop (1980), Stika and Morrison (1981), Horwitz (1982), Iyengar (1989), Smoley (1992), and Smith (1994), factors that require sample preparation, detection limits for the element(s) of interest, and accuracy and precision considerations (LaFluer, 1976). A list of trace element detection limits by atomic optical spectroscopic methods is given in Table 6.1.

TABLE 6.1 Comparison of aqueous detection limits for atomic optical spectroscopic methods

Element	ng/L				
	FAAL	FAFL	FAE	RFICP	ETA-AA
Aluminum (Al)	100	100	3	0.2	0.1
Antimony (Sb)	30	50	600	200	0.5
Arsenic (As)	30	100	10,000	40	0. 8
Barium (Ba)	20	—	1	0.02	0.6
Beryllium (Be)	2	10	1,000	0.4	0.003
Bismuth (Bi)	50	5	20,000	50	0.4
Boron (B)	2,500	—	50	5	20

TABLE 6.1 **Comparison of aqueous detection limits for atomic optical spectroscopic methods (continued)**

	ng/L				
Element	*FAAL*	*FAFL*	*FAE*	*RFICP*	*ETA-AA*
Cadmium (Cd)	1	0.001	800	2	0.008
Cerium (Ce)	—	—	10,000	2	—
Cesium (Cs)	50	—	600	—	0.04
Chromium (Cr)	2	5	2	0.3	0.2
Cobalt (Co)	2	5	30	3	0.2
Copper (Cu)	4	0. 5	10	0.1	0.06
Dysprosium (Dy)	200	—	50	4	—
Erbium (Er)	100	—	70	1	—
Europium (Eu)	40	—	0.2	1	0. 5
Gadolinium (Gd)	4,000	—	70	7	—
Gallium (Ga)	50	10	60	0.6	0.1
Germanium (Ge)	100	100	400	4	0.3
Gold (Au)	20	3	2,000	40	0.1
Hafnium (Hf)	—	—	20,000	10	—
Holmium (Ho)	100	—	100	10	—
Indium (In)	30	100	0.4	30	0.04
Iridium (Ir)	1,000	—	3,000	—	—
Iron (Fe)	4	8	5	0.3	1
Lanthanum (La)	2,000	—	1.0	0.4	—
Lead (Pb)	10	10	100	2	0.2
Lithium (Li)	1	—	0.02	0.3	0. 3
Lutetium (Lu)	3,000	—	1,000	8	—
Manganese (Mn)	0.8	1	1	0.06	0.02
Mercury (Hg)	500	0.2	10,000	1	2
Molybdenum (Mo)	30	500	200	0.2	0.3
Neodymium (Nd)	2,000	—	700	10	—
Nickel (Ni)	5	3	20	0.4	0. 9
Niobium (Nb)	1,000	—	1,000	2	—
Osmium (Os)	400	—	2,000	—	—
Palladium (Pd)	10	40	50	2	0.4
Platinum (Pt)	50	—	4,000	80	1
Praseodymium (Pr)	4,000	—	70	30	—
Rhenium (Re)	600	—	200	—	—
Rhodium (Rh)	20	3,000	30	3	0.11
Rubidium (Rb)	5	—	8	—	0.1
Ruthenium (Ru)	60	5,000	300	—	—
Samarium (Sm)	600	—	200	20	—
Scandium (Sc)	100	—	800	3	6

TABLE 6.1   Comparison of aqueous detection limits for atomic optical spectroscopic methods (continued)

Element	ng/L				
	FAAL	FAFL	FAE	RFICP	ETA-AA
Selenium (Se)	100	40	100,000	30	0.9
Silicon (Si)	100	600	3,000	10	0.005
Silver (Ag)	1	0.1	2	4	0.01
Strontium (Sr)	5	30	0.2	0.02	0.1
Tantalum (Ta)	3,000	—	4,000	30	—
Tellurium (Te)	50	5	2,000	80	0.1
Terbium (Tb)	2,000	—	30	20	—
Thallium (Tl)	20	8	20	200	1
Thorium (Th)	—	—	10,000	3	0.9
Thulium (Tm)	40	—	4	7	—
Tin (Sn)	50	50	100	30	20
Titanium (Ti)	100	4,000	30	0.2	4
Tungsten (W)	3,000	—	600	1	—
Uranium (U)	20,000	—	5,000	30	—
Vanadium (V)	20	70	7	0.2	0.3
Ytterbium (Yb)	20	—	0.2	0.04	0.07
Yttrium (Y)	300	—	30	0.06	—
Zinc (Zn)	1	0.02	10,000	2	0.003
Zirconium (Zr)	4,000	—	5,000	0.4	—

Key to symbols: FAAL = flame atomic absorption with line source; FAFL = flame atomic fluorescence with line source; FAE = flame atomic emission; RFICP = radiofrequency inductively coupled plasma; ETA-AA = electrothermal atomization–atomic absorption.

MULTIELEMENT ANALYSIS

With most of the analytical procedures in use today, a suite of elements can be determined in one analytical step. For example, Nielsen (1986) reported on antimony, barium, boron, bromine, cesium, germanium, rubidium, silver, strontium, tin, titanium, zirconium, beryllium, bismuth, gallium, gold, indium, niobium, scandium, tellurium, thallium, and tungsten, which have been detected in various types of organisms based on assays conducted using these new analytical techniques. Tsalev (1984) not only describes in detail the AAS method of analyzing biological samples, but provides assay results for many body parts and fluids for aluminum, antimony, arsenic, barium, beryllium, bismuth, boron, cadmium, chromium, cobalt, copper, gallium, germanium, gold, indium, iron, lead, lithium, manganese, mercury, molybdenum, nickel, palladium, platinum, rubidium, selenium, silicon, silver, strontium, tellurium, thallium, tin, vanadium, and zinc. Similar examples can be given for many of the trace elements that can be determined in a wide variety of substances (Walsh, 1971; Allen, 1974; Facchetti, 1983; Boumans, 1984; Van Loon, 1982, 1985; Sansoni, 1987; Iyengar, 1989; Barnes, 1991; A.A. Jones, 1991; Smoley, 1992; Grosser, 1995).

With many of the analytical limitations that existed a few years ago overcome, ecologists, biologists, and physiologists now have access to the analytical tools necessary to assay most any substance for its trace element content at concentrations below nanogram levels. Therefore, much of the focus today by ecologists, biologists, and physiologists is on sampling (Gomez et al., 1986; Keith, 1988; Markert, 1994c, 1995a) and sample preparation (Gorsuch, 1970, 1976; Bock, 1978) required prior to analysis.

ANALYTICAL REQUIREMENTS

With the increasing complexity of analytical procedures and instrumentation, much of the analytical work done today for ecologists, biologists, and physiologists is by analytical chemists who may or may not be directly involved in the applied or scientific research being conducted. Therefore, laboratory quality assurance (Berman, 1980; Garfield, 1984; Dux, 1986; Taylor, 1987; Markert, 1995b) and quality control (Anon., 1979b, 1993; Fischbeck, 1980; Hanlon, 1996) are becoming major issues as the volume of analytical data entering the literature increases, together with concerns for its accuracy. Analytical laboratories are having to tighten their procedures to more carefully control the laboratory environment (Kammin et al., 1995) and need to use only high-quality pure reagents in order to minimize the introduction of extraneous elements into prepared samples prior to the assay.

The standardization of methods of analysis is also becoming of major importance, with the U.S. Environmental Protection Agency (Anon., 1983), the U.S. Geological Survey (Fishman and Friedman, 1985), and the Association of Official Analytical Chemists (Helrich, 1995) as the lead agencies in establishing recognized standard methods of analysis. The Soil and Plant Analysis Council has also published the *Handbook on Reference Methods for Soil Analysis* (Jones, 1992).

STANDARD REFERENCE MATERIALS

Another significant advancement has been the increased availability of reference standards for verification of the accuracy of a determination. A number of agencies, both nationally and internationally, are continuing to provide and expand their supply and variety of reference standards. The lead organization in the United States is the National Institute of Standards and Technology, Gaithersburg, MD (Standard Reference Materials Catalog, 1995–96, NIST Special Publication 260). Taylor (1985) has written a bulletin on how standard reference materials are used as an integral part of quality assurance/good laboratory practice (Anon., 1979a, b; Garfield, 1984; Dux, 1986; Taylor, 1987; Anon., 1993; Quevanviller, 1996).

REFERENCES

Adriano, D.C. 1986. *Trace Elements in the Terrestrial Environment.* Springer-Verlag, New York.

Alejar, A.A., R.M. Macandog, J.R. Velasco, and Z.N. Sierra. 1988. Effects of lanthanum, cerium, and chromium on germination and growth of some vegetable species. *Philipp. Agric.* 71(2):185–197.

Allen, H.E., E.M. Perdue, and D. Brown. 1993. *Metals in Groundwater.* Lewis Publishers, Boca Raton, FL.

Allen, S.E. (ed.). 1974. *Chemical Analysis of Ecological Materials.* Blackwell Scientific Publications, Oxford, England.

Alloway, B.J. (ed.). 1995. *Heavy Metals in Soils,* 2nd edition. Blackie Academic and Professional, London, England.

Amann, B.T., P. Mulqeen, and W.D.W. Horrocks. 1992. A continuous spectrophotometric assay for the activation of plant NAD kinase by calmodulin, calcium(II), and europium(III) ions. *J. Biochem. Biophys. Meth.* 25:207–217.

Andersen, M.A., J. Zorrilla-Rios, C.W. Horn, M.J. Ford, and R.W. McNew. 1985. Chromium and ytterbium as single markers for estimating duodenal digesta flow of steers grazing wheat pasture. *Okla. Agr. Exp. Stn. M.P.* 117:161–164.

Anderson, R.R. 1992. Comparison of trace elements in milk of four species. *Am. Dairy Sci. Assoc.* 75:3050–3055.

Anon. 1979a. Nonclinical Laboratory Studies, Good Laboratory Practices. *Federal Register,* Part II, Volume 43, No. 247. Department of Health, Education and Welfare. U.S. Food and Drug Administration, Washington, DC.

199

Anon. 1979b. Handbook for Analytical Quality Control in Water and Wastewater Laboratories. EPA-600/4-79-019. U.S. Environmental Protection Agency, Environmental Monitoring and Support Laboratory, Cincinnati, OH.

Anon. 1983. Methods for Chemical Analysis of Water and Wastes. EPA-600/4-79-020. U.S. Environmental Protection Agency, Environmental Monitoring and Support Laboratory, Cincinnati, OH.

Anon. 1993. Good Laboratory Practice Standards: Inspection Manual. EPA 723-B-93-001. U.S. Environmental Protection Agency, Office of Compliance Monitoring (7204W), Washington, DC.

Arambel, M.J., E.E. Bartley, J.F. Higginbotham, G.G. Simons, and L.J. Ruzback. 1986. Neutron activation analysis of cerium, chromium, lanthanum and samarium in duodenal digesta. *Nutr. Rep. Int.* 33:861–868.

Arnon, D.L. and P.R. Stout. 1939. The essentiality of some elements in minute quantity for plants with special reference to copper. *Plant Physiol.* 14:371–375.

Asher, C.J. 1991. Beneficial elements, functional nutrients, and possible new essential elements, pp. 703–723. In: J.J. Mortvedt et al. (eds.). *Micronutrients in Agriculture,* 2nd edition. SSSA Book Series No. 4. Soil Science Society of America, Madison, WI.

Ashworth, W. 1991. *The Encyclopedia of Environmental Studies,* p. 397. Facts On File, New York.

Baker, D.L. 1971. A new approach to soil testing. *Soil Sci.* 112:381–391.

Baker, D.L. 1973. A new approach to soil testing. II. Ionic equilibria involving H, K, Ca, Mg, Mn, Fe, Cu, Zn, Na, P, and S. *Soil Sci. Soc. Am. Proc.* 27:537–541.

Baker, D.L. 1985. Chemical monitoring of soils for environmental quality, pp. 3–23. In: D.L. Baker and M.R. Murray (eds.). *Application of Soil Environmental Science in Landspreading and Monitoring of Toxic Chemicals.* Institute for Research on Land and Water Resources, The Pennsylvania State University, University Park.

Baker, D.L. and M.C. Amacher. 1981. Development and Interpretation of a Diagnostic Soil Testing Program. Pennsylvania Agricultural Experiment Station Bulletin 826. University Park.

Baker, D.L. and N.H. Suhr. 1982. Atomic absorption and flame emission spectrometry, pp. 13–27. In: A.L. Page (ed.). *Methods of Soil Analysis, Part 2. Chemical and Microbiological Properties.* Agronomy Monograph No. 9. American Society of Agronomy, Madison, WI.

Barnes, R.M. (ed.). 1978. *Applications of Inductively Coupled Plasmas to Emission Spectroscopy.* The Franklin Institute Press, Philadelphia, PA.

Barnes, R.M. 1991. *Developments in Atomic Plasma Spectrochemical Analysis.* Heyden, London, England.

Beckett, P.H.T. and R.D. Davis. 1977. Upper critical levels of toxic elements in plants. *New Phytol.* 79:95–106.

Bederka, J.P., T.M. Lueken, S. Brudno, and R.S. Waters. 1985. Elemental balances in the human, pp. 304–313. In: D.D. Hemphill (ed.). *Trace Substances in Environmental Health—XIX.* University of Missouri, Columbia.

Berger, K.C. and E. Truog. 1939. Boron determination in soils and plants. *Ind. Eng. Chem. Anal. Ed.* 11:540–545.

Bergmann, W. 1983. *Ernährunggstörungen bei Kulturpflanzen, Entstehung und Diagnose.* Gustav Fischer Verlag, Jena, Germany.

Bergmann, W. and P. Neubert. 1976. *Pflanzendiagnose und Pflanzenanalyse.* Fischer, Jena, Germany.

Berman, G.A. 1980. Testing Laboratory Performance: Evaluation and Accreditation. National Bureau of Standards, Washington, DC.

Berrow, M.L. and J.C. Burridge. 1979. Sources and distribution of trace elements in soils and related crops, pp. 206–209. In: *Int. Conf. Manage. Contr. Heavy Metals Environ.* London, England.

Bingham, F.T., F.J. Perya, and W.M. Jarrell. 1986. Metal toxicity to agricultural crops. *Metal. Ion Biol. Sys.* 20:119–156.

Bock, R. 1978. *Handbook of Decomposition Methods in Analytical Chemistry.* International Textbook Company, Glasgow, Scotland.

Boon, D.Y. and P.N. Soltanpour. 1991. Estimating total Pb, Cd, and Zn in contaminated soils from NH_4HCO_3-DTPA extractable levels. *Commun. Soil Sci. Plant Anal.* 22:369–378.

Boumans, P.W.J.M. (ed.). 1984. *Inductively Coupled Plasma Emission Spectroscopy. Part II: Applications and Fundamentals.* John Wiley and Sons, New York.

Brown, P.H., R.M. Welsh, and E.E. Cary. 1987. Nickel: A micronutrient essential for higher plants. *Plant Physiol.* 85:801–803.

Brown, P.H., A.H. Rathjen, R.D. Graham, and D.E. Tribe. 1990. Rare-earth elements in biological systems, pp. 433–452. In: K.A. Gschneidner, Jr. and L. Eyring (eds.). *Handbook on the Physics and Chemistry of Rare-Earths,* Volume 13. Elsevier Science Publishers, Amsterdam, The Netherlands.

Chaney, R.L. 1980. Health risks associated with toxic metals in municipal sludge, pp. 59–67. In: G. Bitton, B.L. Damron, G.T. Edds, and J.M. Davidson (eds.). *Health Risks of Land Application.* Ann Arbor Science, Ann Arbor, MI.

Chaney, R.L. 1983. Plant uptake of inorganic waste constituents, pp. 50–76. In: J.F. Parr, P.B. Marsh, and J.M. Kla (eds.). *Land Treatment of Hazardous Wastes.* Noyes Data Corporation, Park Ridge, NJ.

Chaney, R.L., J.F. Bruins, D.E. Baker, R.F. Korcak, J.E. Smith, and D. Cole. 1987. Transfer of sludge-applied trace elements to the food chain, pp. 67–93. In: A.L. Page, T. Logan, and J. Ryan (eds.). *Land Application of Sludge.* Lewis Publishers, Chelsea, MI.

Collys, K., R. Cleymaet, D. Slop, E. Quartier, and D. Coomans. 1992. The influence of lanthanum on fluoride uptake by sound and surface-softened bovine enamel *in vitro. Trace Elem. Med.* 9:97–101.

Conner, J. and S.T. Bass. 1959. The determination of suitable circuit parameters for a high voltage spark discharge source in the spectrochemical analysis of plant material. *Appl. Spectrosc.* 16:150–155.

Cox, F.R. and E.J. Kamprath. 1972. Micronutrient soil tests, pp. 289–317. In: J.J. Mortvedt, P.N. Giordano, and W.L. Lindsay (eds.). *Micronutrients in Agriculture.* Soil Science Society of America, Madison, WI.

Danford, D.E. 1989. Clinical aspects of trace element, pp. 173–179. In: C. Chazot, M. Abdulla, and P. Arnaud (eds.). *Current Trends in Trace Elements Research.* Smith-Gordon, New York.

Davis, S.N. and DeWiest, J.M. 1966. *Hydrogeology.* Interscience, New York.

Davis, R.D., P.H. Beckett, and E. Wollan. 1978. Critical levels of twenty potentially toxic elements in young spring barley. *Plant Soil* 49:395–408.

Dekock, P.C. and R.L. Mitchell. 1957. Uptake of chelated metals by plants. *Soil Sci.* 84:55–62.

De Temmerman, L.O., H. Hoenig, and P.O. Scokart. 1984. Determination of "normal" levels and upper limit values of trace elements in soils. *Z. Pflanzen. Bodenk.* 147:687–694.

Diatloff, E., F.W. Smith, and C.J. Asher. 1995. Rare-earth elements and plant growth. I. Effects of lanthanum and cerium on root elongation of corn and mungbean. *J. Plant Nutr.* 18(10):1963–1976.

Dux, J.P. 1986. *Handbook of Quality Assurance for the Analytical Chemistry Laboratory.* van Nostrand Reinhold, New York.

Emsley, John. 1991. *The Elements,* 2nd edition. Clarendon Press, Oxford, England.

Epstein, E. 1965. Mineral nutrition, pp. 438–466. In: J. Bonner and J.E. Varner (eds.). *Plant Biochemistry.* Academic Press, Orlando, FL.

Epstein, E. 1972. *Mineral Nutrition of Plants: Principles and Perspectives.* John Wiley and Sons, New York.

Eskew, D.L., R.M. Welch, and E.E. Cary. 1983. Nickel: An essential micronutrient for legumes and possibly all higher plants. *Science* 222:621–623.

Facchetti, S. (ed.). 1983. *Analytical Techniques for Heavy Metals in Biological Fluids.* Elsevier, Amsterdam, The Netherlands.

Faelten, S. 1981. *The Complete Book of Minerals for Health.* Rodale Press, Emmaus, PA.

Fischbeck, R.D. 1980. Good laboratory practices. *Am. Lab.* 12:125–130.

Fishman, M.J. and L.C. Friedman (eds.). 1985. Methods for Determination of Inorganic Substances in Water and Fluvial Sediments. Techniques of Water-Resources Investigations of the United States Geological Survey. Book 5, Chapter A1. U.S. Geological Survey, Denver, CO.

Fodor P. and Molnár, E. 1993. Honey as an environmental indicator: Effect of sample preparation on trace element determination by ICP-AES. *Mikrochim. Acta* 112: 113–116.

Foy, C.D., R.L. Chaney, and M.C. White. 1978. The physiology of metal toxicity in plants. *Annu. Rev. Plant Physiol.* 29:511–566.

Frieden, E. (ed.). 1981. *Biochemistry of the Essential Ultratrace Elements.* Plenum Press, New York.

Garfield, F.M. 1984. *Quality Assurance Principles for Analytical Laboratories.* Association of Official Analytical Chemists, Arlington, VA.

Gebhardt, S.E. and J.M. Holden. 1992. Provisional Table on the Selenium Content of Foods. Human Nutritional Information Service HNIS/PT-109. Agricultural Research Service, U.S. Department of Agriculture, Washington, DC.

Gebhardt, S.E., R. Cutrufelli, and R.H. Mathews. 1982. Composition of Foods, Fruits and Fruit Juices: Raw, Processed, Prepared. Agricultural Handbook No. 8–9. U.S. Department of Agriculture, Human Nutrition Information Service, Washington, DC.

Gibson, R.S. 1990. Essential trace elements and their nutritional importance in the 1990s. *J. Can. Diet Assoc.* 51:292–296.

Glass, A.D.M. 1989. *Plant Nutrition: An Introduction to Current Concepts.* Jones and Bartlett Publishers, Boston, MA.

Gomez, A., R. Leschber, and P. L'Hermite (eds.). 1986. *Sampling Problems for the Chemical Analysis of Sludge, Soils, and Plants.* Elsevier Applied Science Publishers, New York.

Gorsuch, T.T. 1970. *Destruction of Organic Matter,* Volume 39. International Series Monograph. Analytical Chemistry. Pergamon Press, New York.

Gorsuch, T.T. 1976. Dissolution of organic matter, pp. 491–508. In: P.D. LaFluer (ed.). Accuracy in Trace Analysis: Sampling, Sample Handling, Analysis, Volume 1. National Bureau of Standards Special Publication 422, Washington, DC.

Grosser, Z.A. 1995. Inorganic methods update. *Environ. Test. Anal.* 4(3):38–45.

Günter, K., A. von Bohlen, G. Paprott, and R. Klockenhämper. 1992. Multielement analysis of biological reference materials by total-reflection X-ray fluorescence and inductively coupled plasma mass spectrometry in the semiquant mode. *Fresenius J. Anal. Chem.* 342:444–448.

Guo, B. 1985. Present and future situation of rare-earth research in Chinese agronomy, pp. 1522–1526. In: Proceedings International Conference. Beijing, China.

Guo, B. 1987. A new application of rare-earth—agriculture, pp. 237–246. In: *Rare-Earth Horizons.* Australian Department of Industry and Commerce, Canberra, Australia.

Gupta, U.C. 1993. Boron, molybdenum, and selenium, pp. 91–99. In: M.R. Carter (ed.). *Soil Sampling and Methods of Analysis.* Lewis Publishers, Boca Raton, FL.

Gupta, U.C. and J.H. Watkinson. 1985. Agricultural significance of selenium. *Outlook Agric.* 14:183–189.

Halliday, D.J. and M.E. Trenkel (eds.). 1992. *IFA World Fertilizer Use Manual.* International Fertilizer Industry Association, Paris, France.

Hamilton, E.I. and M.J. Minski. 1973. Abundance of the chemical elements in man's diet and possible reactions with environmental factors. *Sci. Total Environ.* 1:375–394.

Hanlon, E. 1996. Laboratory quality: A method for change. *Commun. Soil Sci. Plant Anal.* 27(3/4):307–325.

Haskin, L.A. and K.E. Ziege. 1971. Neutron activation: Techniques and possible uses in soil and plant analysis, pp. 185–209. In: L.M. Walsh (ed.). *Instrumental Methods for Analysis of Soils and Plant Tissue.* Soil Science Society of America, Madison, WI.

Helmke, P.A. 1982. Neutron activation analysis, pp. 67–84. In: A.L. Page (ed.). *Methods of Soil Analysis, Part 2. Chemical and Microbiological Properties.* Agronomy Monograph No. 9. American Society of Agronomy, Madison, WI.

Helrich, K. (ed.). 1995. *Official Methods of Analysis of the Association of Official Analytical Chemists: Agricultural Chemicals, Contaminants, Drugs,* Volume 1. Association of Official Analytical Chemists, Arlington, VA.

Hewitt, E.J. 1966. Sand and Water Culture Methods Used in the Study of Plant Nutrition. Technical Communication No. 22 (revised). Commonwealth Bureau of Horticulture and Plantation Crops, East Malling, Maidstone, Kent, England.

Hislop, J.S. 1980. Choice of the analytical method, pp. 747–767. In: P. Brätter and P. Schramel (eds.). *Trace Element Analytical Chemistry in Medicine and Biology.* DeGruyter, Berlin, Germany.

Horovitz, C.T. 1988. Is the major part of the periodic system really essential for life? *J. Trace Elem. Elect. Health Dis.* 2:135–144.

Horwitz, W. 1982. Evaluation of analytical method used for regulation of foods and drugs. *Anal. Chem.* 54(1):67A–76A.

Huiliang, H., D. Jagner, and L. Renman. 1987. Simultaneous determination of mercury(II), copper(II), and bismuth(III) in urine by flow constant-current stripping analysis with a gold fibre electrode. *Anal. Chim. Acta* 202:117–122.

Ichihashi, H., H. Morita, and R. Tatsukawa. 1992. Rare-earth elements (REEs) in naturally grown plants in relation to their variation in soils. *Environ. Pollut.* 76:157–162.

Iyengar, G.V. 1989. *Elemental Analysis of Biological Systems, Biological, Medical, Environmental: Composition and Methodological Aspects.* CRC Press, Boca Raton, FL.

Jancsó, N. 1961. Inflammation and the inflammatory mechanisms. 3. *J. Pharm. Pharmacol.* 13:577–594.

Jenkins, D.W. 1981. Biological Monitoring of Toxic Trace Elements. EPA-600/S3-80-090. U.S. Environmental Protection Agency, Environmental Monitoring Systems Laboratory, Las Vegas, NV.

Johnson, G.V. and P.E. Fixen. 1990. Testing soils for sulfur, boron, molybdenum, and chlorine, pp. 265–273. In: R.L. Westerman (ed.). *Soil Testing and Plant Analysis.* SSSA Book Series No. 3. Soil Science Society of America, Madison, WI.

Jones, A.A. 1982. X-ray fluorescence spectrometry, pp. 85–121. In: A.L. Page (ed.). *Methods of Soil Analysis, Part 2. Chemical and Microbiological Properties.* Agronomy Monograph No. 9. American Society of Agronomy, Madison, WI.

Jones, A.A. 1991. X-ray fluorescence spectrometry, pp. 287–324. In: K.A. Smith (ed.). *Soil Analysis: Modern Instrumental Techniques,* 2nd edition. Marcel Dekker, New York

Jones, J.B. Jr. (ed.). 1980. *Handbook on Reference Methods for Soil Analysis.* Soil and Plant Analysis Council, Athens, GA.

Jones, J.B. Jr. 1989. Elemental content of Hungarian cropland soil using the Mehlich No. 3 extracting reagent and an ICP spectrometer. *Acta Agron. Hung.* 38:201–209.

Jones, J.B. Jr. 1991. Plant tissue analysis in micronutrients, pp. 477–521. In: J.J. Mortvedt (ed.). *Micronutrients in Agriculture.* SSSA Book Series No. 4. Soil Science Society of America, Madison, WI.

Jones, J.B. Jr. (ed.). 1992. *Handbook on Reference Methods for Soil Analysis—Soil and Plant Analysis Council.* St. Lucie Press, Delray Beach, FL.

Kabata-Pendias, A. and H. Pendias. 1984. *Trace Elements in Soils and Plants.* CRC Press, Boca Raton, FL.

Kabata-Pendias, A. and H. Pendias. 1994. *Trace Elements in Soils and Plants,* 2nd edition. CRC Press, Boca Raton, FL.

Kádas, I. and K. Jobst. 1973. Liver damage induced by lanthanum trichloride. *Acta Morph. Acad. Sci. Hung.* 21:27–37.

Kammin, W.R., S. Cull, R. Knox, J. Ross, M. McIntosh, and D. Thomson. 1995. Labware cleaning protocols for the determination of low-level metals by ICP-MS. *Am. Environ. Lab.* 11/95–12/95:1, 6, 8.

Keith, L.H. (ed.). 1988. *Principles of Environmental Sampling. ECS Professional Reference Book.* American Chemical Society, Washington, DC.

Kubota, J. and V.A. Lazar. 1971. X-ray emission spectrography: Techniques and uses for plant and soil studies, pp. 67–83. In: L.M. Walsh (ed.). *Instrumental Methods for Analysis of Soils and Plant Tissue.* Soil Science Society of America, Madison, WI.

LaFluer, P.D. (ed.). 1976. Accuracy in Trace Analysis: Sampling, Sample Handling, Analysis, Volume 1. National Bureau of Standards Special Publication 422, Washington, DC.

Lakanen, E. and R. Ervio. 1971. A comparison of eight extractants for the determination of plant available micronutrients in soils. *Suom. Maataloustiet. Seuran Julk.* 123: 232–233.

Liang, J. and R.E. Karamanos. 1993. DTPA-extractable Fe, Mn, Cu, and Zn, pp. 87–99. In: M.R. Carter (ed.). *Soil Sampling and Methods of Analysis.* Lewis Publishers, Boca Raton, FL.

Lindsay, W.L. and W.A. Norvell. 1978. Development of a DTPA soil test for zinc, iron, manganese and copper. *Soil Sci. Soc. Am. J.* 42:421–428.

Liu, Z. 1988. The effects of rare-earth elements on growth of crops, pp. 23–25. In: I. Pais (ed.). Proceedings 3rd International Trace Element Symposium. Budapest, Hungary.

Logan, T.J. and R.L. Chaney. 1983. Utilization of municipal wastewater and sludge on land—metals, pp. 235–326. In: A.L. Page, T.L. Gleason, III, J.E. Smith, Jr., I.K. Iskandar, and L.E. Sommers (eds.). *Proceedings of the 1983 Workshop on Utilization of Municipal Wastewater and Sludge on Land.* University of California, Riverside.

Logan, T.J. and S.J. Traina. 1993. Trace metals in agricultural soils, pp. 309–347. In: H.E. Allen, E.M. Perdue, and D. Brown (eds.). *Metals in Groundwater.* Lewis Publishers, Boca Raton, FL.

MacNicol, R.D. and P.H.T. Beckett. 1985. Critical tissue concentrations of potentially toxic elements. *Plant Soil* 85:107–129.

Markert, B. 1992. Presence and significance of naturally occurring chemical elements of the periodic system in the plant organism and consequences for future investigations on inorganic environmental chemistry in ecosystems. *Vegetation* 103:1–30.

Markert, B. 1993. Interelement correlations detectable in plant samples based on data from reference materials and highly accurate research samples. *Fresnius J. Anal. Chem.* 345:318–322.

Markert, B. 1994a. Plants as biomonitors—Potential advantages and problems, pp. 601–613. In: D.C. Adriano, Z.S. Chen, and S.S. Yang (eds.). *Biogeochemistry of Trace Elements.* Science and Technology Letters, Northwood, NY.

Markert, B. 1994b. *Element Concentration Cadasters in Ecosystems.* International Institute of Advanced Ecological and Economic Studies, Zittau, Germany.

Markert, B. 1994c. *Environmental Sampling for Trace Analysis.* VCH Publishers, Deerfield Beach, FL.

Markert, B. 1995a. *Instrumental Multielement Analysis in Plant Materials—A Modern Method in Environmental Chemistry and Tropical Systems Research.* Technologia Ambiental. MCT, CNpq CETEM. Rio de Janeiro, Brazil.

Markert, B. 1995b. Quality assurance of plant sampling and storage, pp. 215–254. In: Ph. Quevanviller (ed.). *Quality Assurance in Environmental Monitoring—Sampling and Sample Pretreatment.* VCH Publishers, New York.

Markert, B. and W. Geller. 1994. Multielement analysis of tropical lakes, pp. 27–43. In: R.M. Pinto-Coelho, A. Giana, and E. von Sperling (eds.). *Ecology and Human Impact on Lakes and Reservoirs in Minas Gerais with Special Reference to Future Development and Management Strategies.* SEGRAC-Belo-Horizonte (MG), Brazil.

Markert, B., U. Reus, and U. Herpin. 1994. The application of TXRF in instrumental multielement analysis of plants, demonstrated with species of moss. *Sci. Total Environ.* 152:213–230.

Martens, D.C. and W.L. Lindsay. 1990. Testing soils for copper, iron, manganese, and zinc, pp. 229–273. In: R.L. Westerman (ed.). *Soil Testing and Plant Analysis.* SSSA Book Series No. 3. Soil Science Society of America, Madison, WI.

McDowell, L.R. 1989. Soil-plant-animal-man relationships, pp. 23–26. In: C. Chazot, M. Abdulla, and P. Arnaud (eds.). *Current Trends in Trace Elements Research.* Smith-Gordon, New York.

McDowell, L.R. 1992. *Minerals in Animal and Human Nutrition.* Academic Press, New York.

McGrath, S.T., J.R. Sanders, and T.M. Adams. 1985. Comparison of soil solution and chemical extractants to estimate metal availability to plants. *J. Sci. Food Agric.* 36:532–533.

Meehan, B., K. Peverill, and A. Skroce. 1993. The impact of bioavailable rare-earth elements in Australia agricultural soils, p. 36. In: Australian Soil and Plant Analysis First National Workshop on Soil and Plant Analysis. Ballarat, Australia.

Meloni, S. and N. Genova. 1987. Rare-earth elements in the NBS standard reference materials: Spinach, orchard leaves, pine needles and bovine liver. *Sci. Total Environ.* 64:13–20.

Melsted, S.W. 1973. Soil-plant relationships (some practical considerations in waste management). In: *Proceedings Joint Conference on Recycling Municipal Sludges and Effluents on Land.* University of Illinois, Urbana.

Mertz, W. 1980. Implications of the new trace elements for human health, pp. 11–15. In: M. Anke et al. (eds.). *Proceedings 3rd Spurenelement Symposium.* Jena, Germany.

Mertz, W. 1989. Trace element requirements and current recommendations, pp. 1–6. In: C. Chazot, M. Abdulla, and P. Arnaud (eds.). *Current Trends in Trace Elements Research.* Smith-Gordon, New York.

Metcalfe, E. 1987. *Atomic Absorption and Emission Spectrometry.* John Wiley and Sons, New York.

Miller, E.R., X. Liei, and D.E. Ullrey. 1991. Trace elements in animal nutrition, pp. 593–662. In: J.J. Mortvedt et al. (eds.). *Micronutrients in Agriculture,* 2nd edition. SSSA Book Series No. 4. Soil Science Society of America, Madison, WI.

Mitchell, R.L. 1964. The Spectrochemical Analysis of Soils, Plants and Related Materials. Technical Communication No. 44A. Commonwealth Bureau of Soils, Harpenden. Commonwealth Agricultural Bureaux, Farnham Royal, Bucks, England.

Montaser, A. and D.W. Golightly (eds.). 1987. *Inductively Coupled Plasmas in Analytical Atomic Spectrometry.* VCH Publishers, New York.

Morrison, G.H. 1979. Elemental trace analysis of biological materials. *Crit. Rev. Anal. Chem.* 8:287–320.

Nagy, I., I. Kadas, and K. Jobst. 1976. Lanthanum trichloride induced blood coagulation defect and liver injury. *Hematology* 10:353–359.

Nieboer, E. and W.E. Sanford. 1985. Essential, toxic, and therapeutic functions of metals (including determinants of reactivity), pp. 205–245. In: E. Hodgen, J.R. Bend, and R.M. Philpot (eds.). *Reviews in Biochemical Toxicity.* Elsevier Publishing, New York.

Nielsen, F.H. 1984. Ultratrace elements in nutrition. *Annu. Rev. Nutr.* 4:21–41.

Nielsen, F.H. 1986. Other elements: Sb, Ba, B, Br, Cs, Ge, Rb, Ag, Sr, Sn, Ti, Zr, Be, Bi, Ga, Au, In, Nb, Sc, Te, Tl, W, pp. 415–463. In: W. Mertz (ed.). *Trace Elements in Human and Animal Nutrition,* Volume 2, 5th edition. Academic Press, New York.

Oldfied, J.E. (ed.). 1994. Risks and Benefits of Selenium in Agriculture. CAST Issue Paper Number 3. Council for Agricultural Science and Technology, Ames, IA.

Page, A.L., T. Logan, and J. Ryan (eds.). 1987. *Land Application of Sludge.* Lewis Publishers, Chelsea, MI.

Pais, I. 1989. Developments in the trace element research: Our knowledge at present and trends in the near future. *Acta Agron. Hung.* 38(1–2):167–175.

Pais, I. 1991. Criteria of essentiality, beneficiality and toxicity. What is too little and too much? pp. 59–77. In: I. Pais (ed.). *Proceedings of IGBP Symposium,* Budapest, Hungary.

Pais, I. 1992. Criteria of essentiality, beneficiality, and toxicity of chemical elements. *Acta Aliment.* 21(2):145–152.

Pais, I. 1995. Unpublished data.

Pallis, J.E., Jr. and J.B. Jones, Jr. 1978. Platinum uptake by horticultural crops. *Plant Soil* 50:207–212.

Pennington, J.A.T. and B. Young. 1990. Iron, zinc, copper, manganese, selenium, and iodine in foods from the United States total diet study. *J. Food Comp. Anal.* 3: 166–184.

Pennington, J.A.T., S.A. Schoen, G.D. Salmon, B. Young, R.D. Johnson, and R.W. Marts. 1995. Composition of core foods in the U.S. food supply, 1982–1991. *J. Food Comp. Anal.* 8:91–128.

Pietra, R., E. Sabbioni, L. Ubertalli, E. Orvini, C. Vocaturo, F. Colombo, and M. Zanoni. 1985. Trace elements in tissues of a worker affected by rare-earth pneumoconiosis. A study carried out by neutron activation analysis. *J. Radioanal. Nucl. Chem.* 92:247–259.

Piper, C.S. 1942. *Soil and Chemical Analysis.* Hassell Press, Adelaide, Australia.

Porter, J.R. and Lawlor, D.W. (eds.). 1991. *Plant Growth Interactions with Nutrition and Environment.* Society for Experimental Biology. Seminar Series 43. Cambridge University Press, New York.

Prange, A. and H. Schwenke. 1992. Trace element analysis using total reflection X-ray fluorescence spectrometry. *Adv. X-ray Anal.* 35:899–923.

Quevanviller, P. 1996. Certified reference materials for the quality control of total and extractable trace element determinations in soils and sludges. *Commun. Soil Sci. Plant Anal.* 27(3/4):403–418.

Risser, J.A. and D.L. Baker. 1990. Testing soils for toxic metals, pp. 275–298. In: R.L. Westerman (ed.). *Soil Testing and Plant Analysis.* SSSA Book Series No. 3. Soil Science Society of America, Madison, WI.

Robb, D.A. and W.S. Pierpoint. 1983. *Metals and Micronutrients: Uptake and Utilization by Plants.* Academic Press, New York.

Römheld, V. and H. Marschner. 1991. Function of micronutrients in plants, pp. 297–328. In: J.J. Mortvedt et al. (eds.). *Micronutrients in Agriculture,* 2nd edition. SSSA Book Series No. 4. Soil Science Society of America, Madison, WI.

Rump, H.H. and H. Krist. 1988. *Laboratory Manual for the Examination of Water, Waste Water, and Soil.* VCH Publishers, New York.

Salmon, L. and P.A. Cawse. 1991. Instrumental neutron activation analysis, pp. 377–432. In: K.A. Smith (ed.). *Soil Analysis: Modern Instrumental Techniques,* 2nd edition. Marcel Dekker, New York.

Sansoni, B. (ed.). 1987. *Instrumentelle Multielementanalyse.* VCH Verlaggesellschaft, New York.

Schmidt, L.H. 1989. Zur essentialitat der spurenelemente. *Pharmazie* 44:377–380.

Schwarz, K. 1970. Control of the environmental conditions of trace element research: An experimental approach to unrecognized trace element requirements, pp. 25–52. In: C.F. Mills (ed.). *Proceedings International Symposium.* Aberdeen, Scotland.

Shacklette, H.T. 1980. Elements in Fruits and Vegetables from Areas of Commercial Production in the Conterminous United States. Geological Survey Professional Paper 1178. U.S. Government Printing Office, Washington, DC.

Sharp, B.L. 1991. Inductively coupled plasma spectrometry, pp. 63–109. In: K.A. Smith (ed.). *Soil Analysis: Modern Instrumental Techniques,* 2nd edition. Marcel Dekker, New York

Shorrocks, V.M. 1987. Recent developments regarding boron, copper, iron, manganese, molybdenum, selenium and zinc. *Devel. Plant Soil Sci.* 29:270–290.

Shuman, L.M., V.A. Bandel, S.J. Donohue, R.A. Isaac, R.M. Lippert, J.T. Sims, and M.R. Tucker. 1992. Comparison of Mehlich-1 and Mehlich-3 extractable soil boron with hot-water extractable boron. *Commun. Soil Sci. Plant Anal.* 23:1–14.

Sillanpää, M. 1982. *Micronutrients and the Nutrient Status of Soils: A Global Study.* Soil Bulletin 48. FAO, Rome, Italy.

Sillanpää, M. 1990. *Micronutrient Assessment at the Country Level: An International Study.* Soil Bulletin 63. FAO, Rome, Italy.

Sillanpää, M. and H. Jansson. 1992. *Status of Cadmium, Lead, Cobalt, and Selenium in Soils and Plants of Thirty Countries.* FAO Soils Bulletin 65. Rome, Italy.

Sims, J.T. 1991. Recommended soil tests for micronutrients: Manganese, zinc, copper, and boron, pp. 35–40. In: *Recommended Soil Testing Procedures for the Northeastern United States.* Northeastern Regional Publication No. 493. Agricultural Experiment Station, University of Delaware, Newark.

Sims, J.T. and C.V. Johnson. 1991. Micronutrient soil tests, pp. 427–476. In: J.J. Mortvedt et al. (eds.). *Micronutrients in Agriculture.* SSSA Book Series No. 4. Soil Science Society of America, Madison, WI.

Skogerboe, R.K., I.T. Urasa, and G.N. Coleman. 1976. Characterization of a DC plasma as an excitation source for multielement analysis. *Appl. Spectr.* 30:500–507.

Skoryna, S.C., Y. Nagamachi, and V.A. Dvorak. 1989. Changing concepts in essentiality of trace elements in life and health, pp. 21–32. In: D.D. Hemphill (ed.). *Trace Substances in Environmental Health—XXIII*. University of Missouri, Columbia.

Smith, R.-K. 1994. *Handbook of Environmental Analysis,* 2nd edition. Genium Publishing Company, Schenectady, NY.

Smoley, C.K. 1992. *Methods for the Determination of Metals in Environmental Samples*. CRC Press, Boca Raton, FL.

Soltanpour, P.N. and S.M. Workman. 1980. Use of NH_4HCO_3-DTPA soil test to assess availability and toxicity of selenium to alfalfa plants. *Commun. Soil Sci. Plant Anal.* 11:1147–1156.

Soltanpour, P.N., J.B. Jones, Jr., and S.M. Workman. 1982. Optical emission spectrometry, pp. 29–65. In: A.L. Page (ed.). *Methods of Soil Analysis, Part 2. Chemical and Microbiological Properties*. Agronomy Monograph No. 9. American Society of Agronomy, Madison, WI.

Stika, J.W.A. and G.H. Morrison. 1981. Analytical methods for mineral content of human tissues. *Fed. Proc.* 40:2115–2120.

Street, J.J. and W.M. Peterson. 1982. Anodic stripping voltammetry and differential pulse polarography, pp. 133–148. In: A.L. Page (ed.). *Methods of Soil Analysis, Part 2. Chemical and Microbiological Properties*. Agronomy Monograph No. 9. American Society of Agronomy, Madison, WI.

Summerton, A.P. 1992. Incorporation of natural radionuclides and rare-earth elements into a salt-tolerant plant. *J. Radioanal. Nucl. Chem.* 161:421–428.

Swaine, D.J. 1969. The Trace Element Content of Soils. Commonwealth Bureau of Soil Science. Technical Communication No. 46. Commonwealth Agricultural Bureaux, Farnham Royal, Bucks, England.

Takahashi, E., J.F. Ma, and Y. Miyake. 1990. The possibility of silicon as an essential element for higher plants, pp. 99–122. In: *Comments on Agriculture and Food Chemistry*. Gordon and Breach, London.

Taylor, J.K. 1985. Handbook for SRM Users. Special Publication 260-100. U.S. Department of Commerce, National Bureau of Standards, Gaithersburg, MD.

Taylor, J.K. 1987. *Quality Assurance of Chemical Measurements*. Lewis Publishers, Chelsea, MI.

Temmerman, L.O., M. Hoenig, and P.O. Scokart. 1984. Determination of "normal" levels and upper limit values of trace elements in soils. *Z. Pflanzen. Bodenk.* 147:687–694.

Thompson, G. and D. Blanston. 1969. A technique for trace element analysis of powdered material using d.c. arc and photoelectric spectrometry. *Spectrochim. Acta* 24B:335–350.

Thompson, M. and J.N. Walsh. 1983. *A Handbook of Inductively Coupled Plasma Spectrometry.* Blackie, Glasgow, Scotland.

Toro, E., H.A. Das, J.J. Fardy, Z. bin Hamzah, R.K. Iyer, S. Laiyan, N. Leelhaphunt, Y. Muramatsu, R.M. Parr, and I.H. Qurashi. 1994. Toxic heavy metals and other trace elements in foodstuffs from 12 different countries: An IAEA Coordinate Research Program, pp. 415–422. In: *Biological Trace Element Research.* Humana Press, Totowa, NJ.

Tsalev, D.L. 1984. *Atomic Absorption Spectrometry in Occupational and Environmental Health Practice,* Volume II: Determination of Individual Elements. CRC Press, Boca Raton, FL.

Ure, A.M. 1991. Atomic absorption and flame emission spectrometry, pp. 1–62. In: K.A. Smith (ed.). *Soil Analysis: Modern Instrumental Techniques,* 2nd edition. Marcel Dekker, New York.

van Campen, D.R. 1991. Trace elements in human nutrition, pp. 663–701. In: J.J. Mortvedt et al. (eds.). *Micronutrients in Agriculture,* 2nd edition. SSSA Book Series No. 4. Soil Science Society of America, Madison, WI.

van der Leeden, F., L.T. Troise, and D.K. Todd (eds.). 1990. *The Water Encyclopedia.* Lewis Publishers, Boca Raton, FL.

Van Loon, J.C. 1982. *Chemical Analysis of Inorganic Constituents of Water.* CRC Press, Boca Raton, FL.

Van Loon, J.C. 1985. *Selected Methods of Trace Metal Analysis: Biological and Environmental Samples.* John Wiley and Sons, New York.

Varma, Asha. 1991. *CRC Handbook of Inductively Coupled Plasma Emission Spectroscopy.* CRC Press, Boca Raton, FL.

Wallace, A. 1971. *Regulation of the Micronutrient Status of Plants by Chelating Agents and Other Factors.* UCLA 34P51-33. Arthur Wallace, Los Angeles, CA.

Walsh, L.M. (ed.). 1971. *Instrumental Methods for Analysis of Soils and Plant Tissue.* Soil Science Society of America, Madison, WI.

Wang, L., Z. Xu, and R. Tang. 1985. Effects of rare-earth elements on photosynthesis and photofixation of nitrogen in *Anabena azotica,* pp. 1527–1529. In: Proceedings International Conference. Beijing, China.

Wear, J.I. 1965. Boron, pp. 1059–1063. In: C.A. Black (ed.). *Methods of Soil Analysis, Part 2. Chemical and Microbiological Properties,* 2nd edition. Agronomy No. 9. American Society of Agronomy, Madison, WI.

Weiss, G.B. 1974. Cellular pharmacology of lanthanum. *Annu. Rev. Pharmacol.* 14: 343–354.

Welch, R.M., W.H. Allawya, W.A. House, and J. Kubota. 1991. Geographic distribution of trace element problems, pp. 31–57. In: J.J. Mortvedt et al. (eds.). *Micronutrients in Agriculture,* 2nd edition. SSSA Book Series No. 4. Soil Science Society of America, Madison, WI.

Whitney, D.A. 1988. Micronutrient soil tests for zinc, iron, manganese, and copper, pp. 20–22. In: W.C. Dahnke (ed.). *Recommended Chemical Soil Test Procedures for the North Central Region.* North Central Regional Publication No. 221 (revised). North Dakota Agricultural Experiment Station, Fargo.

Wolnik, K.A. F.L. Fricke, S.G. Capar, G.L. Braude, M.W. Meyer, R.D. Satzger, and E. Bonnin. 1983a. Elements in major raw agricultural crops in the United States. 1. Cadmium and lead in lettuce, peanuts, potatoes, soybeans, sweet corn, and wheat. *J. Agric. Food Chem.* 31(6):1240–1244.

Wolnik, K.A. F.L. Fricke, S.G. Capar, G.L. Braude, M.W. Meyer, R.D. Satzger, and R.W. Kuennen. 1983b. Elements in major raw agricultural crops in the United States. 2. Other elements in lettuce, peanuts, potatoes, soybeans, sweet corn, and wheat. *J. Agric. Food Chem.* 31(6):1244–1249.

Wolnik, K.A., F.L. Fricke, S.G. Capar, M.W. Meyer, R.D. Satzger, E. Bonnin, and C.M. Gaston. 1985. Elements in major raw agricultural crops in the United States. 3. Cadmium, lead and other elements in carrots, field corn, onions, rice, spinach, and tomatoes *J. Agric. Food Chem.* 33:807–811.

Ylärnta, T. 1990. The selenium content of some agricultural crops and soils before and after the addition of selenium to fertilizers in Finland. *Ann. Agric. Fenn.* 29: 131–139.

Zarcinas, B.A. 1984. *Analysis of Soil and Plant Material by Inductively Coupled Plasma— Optical Emission Spectrometry.* CSIRO. Division of Soils. Divisional Report No. 70. Commonwealth Scientific and Industrial Research Organization, Australia.

Zhu, Q. and Z. Liu. 1991. Content and distribution of rare-earth elements in soils of China, pp. 194–199. In: S. Portch (ed.). *Proceedings International Symposium on the Role of Sulphur, Magnesium and Micronutrients in Balanced Plant Nutrition.* Chengdu, China.

BOOKS ON THE TRACE ELEMENTS

Adriano, D.C. 1986. *Trace Elements in the Terrestrial Environment.* Springer-Verlag, New York.

Adriano, D.C. and I.L. Brisbin (eds.). 1978. *Environmental Chemistry and Cycling Processes.* NTIS, Springfield, VA.

Adriano, D.C., Z.S. Chen, and S.S. Yang. 1994. *Biogeochemistry of Trace Elements.* Science and Technology Letters, Northwood, NY.

Allen, H.E. and C.-P. Huang. 1995. *Metal Speciation and Contamination of Soil.* Lewis Publishers, Boca Raton, FL.

Allen, H.E., E.M. Perdue, and D. Brown. 1993. *Metals in Groundwater.* Lewis Publishers, Boca Raton, FL.

Alloyway, B.J. (ed.). 1995. *Heavy Metals in Soils,* 2nd edition. Blackie Academic and Professional, London, England.

Ashworth, W. 1991. *The Encyclopedia of Environmental Studies.* Facts On File, New York.

Barber, S.A. 1984. *Soil Nutrient Bioavailability: A Mechanistic Approach.* John Wiley and Sons, New York.

Bennett, W.F. 1993. *Nutrient Deficiencies and Toxicities in Crop Plants.* The American Phytopathological Society, St. Paul, MN.

BoSheng, Guo. 1987. *Rare Earth Horizons.* Australian Department of Industry and Commerce, Canberra, Australia.

Bowen, H.J.M. 1966. *Trace Elements in Biochemistry.* Academic Press, London, England.

Bowen, H.J.M. 1979. *Environmental Chemistry of the Elements.* Academic Press, New York.

Bowen, H.J.M. 1982. *Environmental Chemistry.* Royal Society, London, England.

Browie, S.H.U. and I. Thornton (eds.). 1985. *Environmental Geochemistry and Health.* D. Reidel, Dordrecht, The Netherlands.

CEP Consultants. 1983. *Heavy Metals in the Environment: Heidelberg.* CEP Consultants, Edinburgh, Scotland.

Chapman, H.D. 1966. *Diagnostic Criteria for Plants and Soils.* Division of Agricultural Sciences, University of California, Berkeley.

Chazot, M., M. Abdulla, and P. Arnaud (eds.). 1989. *Current Trends in Trace Elements Research.* Smith-Gordon, New York.

Chen, Y. and Y. Hadar (eds.). 1991. *Iron Nutrition and Interactions in Plants.* Kluwer Academic Publishers, Dordrecht, The Netherlands.

Childers, N.F. (ed.). 1966. *Nutrition of Fruit Crops, Tropical, Sub-Tropical, Temperate Tree and Small Fruits.* Somerset Press, Somerville, NJ.

Davies, B.E. (ed.). 1980. *Applied Soil Trace Elements.* John Wiley and Sons, New York.

Drucker, H. and R.E. Wildung (eds.). 1977. Biological Implications of Metals in the Environment. CONF-750929. NITS, Springfield, VA.

Eisler, R. 1981. *Trace Metal Concentrations in Marine Organisms.* Pergamon Press, New York.

Emsley, J. 1991. *The Elements,* 2nd edition. Clarendon Press, Oxford, England.

Epstein, E. 1972. *Mineral Nutrition of Plants: Principles and Perspectives.* John Wiley and Sons, New York.

Faelten, S. (ed.). 1981. *The Complete Book of Minerals for Health.* Rodale Press, Emmaus, PA.

Frieden, E. (ed.). 1984. *Biochemistry of the Essential Ultratrace Elements.* Plenum, New York.

Gilbert, J. (ed.). 1984. *Analysis of Food Contaminants.* Elsevier Applied Science Publishers, London, England.

Goodhart, R.S. and M.E. Shils (eds.). 1980. *Modern Nutrition in Health and Disease,* Volume 6. Lea and Febiger, Philadelphia, PA.

Gough, L.P., H.T. Shacklette, and A.A. Case. 1979. Element Concentrations Toxic to Plants, Animals, and Man. Geological Survey Bulletin 1466. Department of the Interior, Washington, DC.

Hewitt, E.J. 1966. Sand and Water Culture Methods Used in the Study of Plant Nutrition. Technical Communication No. 22 (revised). Commonwealth Agricultural Bureaux, East Malling, Maidstone, Kent, England.

Jones, J.B. Jr. 1994. *Plant Nutrition Manual.* Micro-Macro International, Athens, GA.

Jones, J.B. Jr. 1994. *Plant Nutrition Basics* (VHS video). St. Lucie Press, Delray Beach, FL.

Jones, J.B. Jr. 1994. *The Micronutrients* (VHS video). St. Lucie Press, Delray Beach, FL.

Jones, J.B. Jr., B. Wolf, and H.A. Mills. 1991. *Plant Analysis Handbook: A Practical Sampling, Preparation, Analysis, and Interpretation Guide.* Micro-Macro International, Athens, GA.

Kabata-Pendias, A. and H. Pendias. 1984. *Trace Elements in Soils and Plants.* CRC Press, Boca Raton, FL.

Kabata-Pendias, A. and H. Pendias. 1994. *Trace Elements in Soils and Plants,* 2nd edition. CRC Press, Boca Raton, FL.

Kitagishi, K. and I. Yamane. 1981. *Heavy Metal Pollution in Soils of Japan.* Japan Scientific Societies Press, Tokyo.

Leeper, G.W. 1978. *Managing the Heavy Metals on the Land.* Marcel Dekker, New York.

Lepp, N.W. (ed.). 1981. *Effect of Heavy Metal Pollution on Plants.* Applied Science Publications, London, England.

Lichen, J. (ed.). 1973. *Manganese.* National Academy of Science, Washington, DC.

Lindsay, W.L. 1979. *Chemical Equilibria in Soils.* John Wiley and Sons, New York.

Loneragan, J.F., A.D. Robson, and R.D. Graham. 1985. *Copper in Soils and Plants.* Academic Press, Sydney, Australia.

Markert, B. (ed.). 1993. *Plants as Biomonitors: Indicators for Heavy Metals in the Terrestrial Environment.* VCH Publishers, New York.

McDowell, L.R. 1992. *Minerals in Animal and Human Nutrition.* Academic Press, New York.

Mertz, W. (ed.). 1986. *Trace Elements in Human and Animal Nutrition,* Volume 2, 5th edition. Academic Press, New York.

Mertz, W. (ed.). 1987. *Trace Elements in Human and Animal Nutrition,* Volume 1, 5th edition. Academic Press, New York.

Mertz, W. and E. Carnatzer (eds.). 1971. *Newer Trace Elements in Nutrition.* Marcel Dekker, New York.

Mills, H.A. and J.B. Jones, Jr. 1996. *Plant Nutrition Manual II.* Micro-Macro Publishing, Athens, GA.

Mortvedt, J.J. (ed.). 1991. *Micronutrients in Agriculture,* 2nd edition. SSSA Book Series No. 4. Soil Science Society of America, Madison, WI.

Mortvedt, J.J., P.M. Giordano, and W.L. Lindsay (eds.). 1972. *Micronutrients in Agriculture.* Soil Science Society of America, Madison, WI.

Nicholas, D.J.D. and A.R. Egan (eds.). 1975. *Trace Elements in Soil-Plant-Animal Systems.* Academic Press, New York.

Nielson, F.H. (ed.). 1977. *Geochemistry and the Environment,* Volume 2. National Academy of Science, Washington, DC.

Nriagu, J.O. (ed.). 1979. *Zinc in the Environment.* John Wiley and Sons, New York.

Nriagu, J.O. (ed.). 1979. *Copper in the Environment.* John Wiley and Sons, New York.

Nriagu, J.O. (ed.). 1980. *Cadmium in the Environment.* John Wiley and Sons, New York.

Nriagu, J.O. (ed.). 1980. *Nickel in the Environment.* John Wiley and Sons, New York.

Nriagu, J.O. (ed.). 1984. *Changing Metal Cycles and Human Health.* Springer-Verlag, Berlin, Germany.

Olson, R.E. (ed.). 1984. *Nutrition Review's Present Knowledge in Nutrition.* The Nutritional Foundation, Washington, DC.

Plucknett, D.L. and H.B. Sprague (eds.). 1989. *Detecting Mineral Nutrient Deficiencies in Tropical and Temperate Crops.* Westview Tropical Agricultural Series No. 7. Westview Press, Boulder, CO.

Prasad, A.S. (ed.). 1993. *Essential and Toxic Elements in Human Health and Disease: An Update.* Wiley-Liss, New York.

Purvis, D. 1985. *Trace Element Contamination of the Environment.* Elsevier Science Publishers, Amsterdam, The Netherlands.

Rechcigl, J.E. 1995. *Soil Amendments and Environmental Quality.* Lewis Publishers, Boca Raton, FL.

Reuter, D.J. and J.B. Robinson (eds.). 1986. *Plant Analysis: An Interpretation Manual.* Inkata Press, Melbourne, Australia.

Robb, D.A. and W.S. Pierpoint (eds.). 1983. *Metals and Micronutrients: Uptake and Utilization by Plants.* Phytochemical Society of Europe Symposia Series No. 21. Academic Press, New York.

Rose, J. 1983. *Trace Elements in Health: A Review of Current Issues.* Butterworths, London, England.

Schrauzer, G.N. and K.F. Klippell (eds.). 1991. *Lithium in Biology and Medicine.* VCH Publishers, New York.

Shaw, A.J. (ed.). 1990. *Heavy Metal Tolerance in Plants: Evolutionary Aspects.* CRC Press, Boca Raton, FL.

Shkolnik, M.Ta. 1984. *Trace Elements in Plants. Developments in Crop Science (6).* Elsevier, Amsterdam, The Netherlands.

Sillanpää, M. 1972. *Trace Elements in Soils and Agriculture.* Soils Bulletin No. 17. FAO, Rome, Italy.

Sillanpää, M. 1990. *Micronutrient Assessment at the Country Level: An International Study.* Soil Bulletin 63. FAO, Rome, Italy.

Sillanpää, M. and H. Jansson. 1992. *Status of Cadmium, Lead, Cobalt, and Selenium in Soils and Plants of Thirty Countries.* Soil Bulletin 65. FAO, Rome, Italy.

Smith, K.T. (ed.). 1988. *Trace Elements in Foods.* Marcel Dekker, New York.

Smith, R.-K. 1994. *Handbook of Environmental Analysis,* 2nd edition. Genium Publishing, Schenectady, NY.

Stoeppler, H. 1992. *Hazardous Metals in the Environment.* Elsevier Publishers, New York.

Swaine, D.J. 1969. The Trace Element Content of Soils. Commonwealth Bureau of Soil Science. Technical Communication No. 46. Commonwealth Agricultural Bureaux, Farnham Royal, Bucks, England.

Tsalev, D.L. 1984. *Atomic Absorption Spectrometry in Occupational and Environmental Health Practice,* Volume II: Determination of Individual Elements. CRC Press, Boca Raton, FL.

Underwood, E.J. 1977. *Trace Elements in Human and Animal Nutrition.* Academic Press, New York.

van der Leeden, F., L.T. Troise, and D.K. Todd (eds.). 1990. *The Water Encyclopedia.* Lewis Publishers, Boca Raton, FL.

Van Loon, J.C. 1985. *Selected Methods of Trace Metal Analysis: Biological and Environmental Samples.* John Wiley and Sons, New York.

Waldron, H.A. (ed.). 1980. *Metals in the Environment.* Academic Press, New York.

Wallace, Arthur. 1971. *Regulation of the Micronutrient Status of Plants by Chelating Agents and Other Factors.* UCLA 34P51-33. Arthur Wallace, Los Angeles.

Welch, R.M. and W.H. Gabelman (eds.). 1984. *Crops as Sources of Nutrients for Humans.* ASA Special Publication 48. American Society of Agronomy, Madison, WI.

INDEX